Ben Stacy Jerrik (Ed.)

Hemus

Ben Stacy Jerrik (Ed.)

Hemus

Botti, Species, Family, Genus, Departments

Part Press

Imprint

Publisher:
Part Press is a trademark of
International Book Market Service Ltd., 17 Rue Meldrum, Beau Bassin, 1713-01 Mauritius
Email: info@bookmarketservice.com
Website: www.bookmarketservice.com

Published in 2013

Printed in: U.S.A., U.K., Germany. This book was not produced in Mauritius.

ISBN: 978-620-1-56337-7

Contents

Articles

References

File_Manager_(Windows)

The "File Manager" is also a development API for the Mac OS

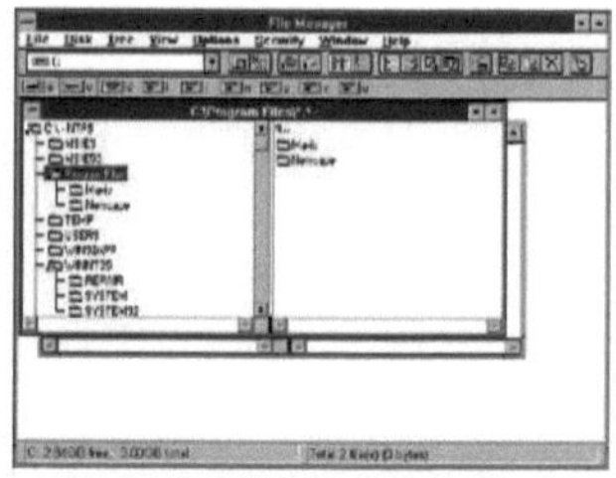

A screenshot of File Manager on Windows NT 3.5 displaying a folder and the contents of drive C:

File Manager is a file manager program bundled with releases of Microsoft Windows between 1990 and 1999.[] It was a single-instance graphical interface, replacing the command-line interface of MS-DOS, to manage files (copy, move, open, delete, search, etc.). Although File Manager was included in Windows 95 and Windows NT 4.0 and some later versions, Windows Explorer was introduced and used as the primary file manager, with file management via a 2-pane view different from File Manager's, and a single-pane view obtained by clicking a "My Computer" icon.

Overview

The program's interface showed a list of directories on the left hand panel, and a list of the current directory's contents on the right hand panel. File Manager allowed a user to create, rename, move, print, copy, search for, and delete files and directories, as well as to set permissions (attributes) such as archive, read-only, hidden or system, and to associate file types with programs. Also available were tools to label and format disks, manage folders for file sharing and to connect and disconnect from a network drive. On Windows NT systems it was also possible to set ACLs on files and folders on NTFS partitions through the shell32 security configuration dialog (also used by Explorer and other Windows file managers). On NTFS drives, individual files or entire folders could be compressed or expanded.

The Windows NT version of File Manager allows users to change directories, files, local, network and user permissions.

From Windows 95 and Windows NT 4.0 onward, File Manager was superseded by Windows Explorer. However, the WINFILE.EXE program file was still included with Windows 95, Windows 98 and Windows Me (16-bit executable), and Windows NT 4.0 (32-bit executable). The last 32-bit WINFILE.EXE build (4.0.1381.318) was distributed as part of Windows NT 4.0 Service Pack 6a (SP6a). The last 16-bit WINFILE.EXE build (4.90.3000) was distributed as part of Windows Me operating system. All versions of File Manager are available at this page [1].

Ian Ellison-Taylor was the shell developer on the Windows 3.1 team responsible for File Manager and Print Manager.[2]

Versions

16-bit

The original version of File Manager was a 16-bit program that supported the 8.3 file names that were in use at the time.

It did not support the extended file names that became available in Windows 95 - including long file names and file names containing spaces. Instead, it would display only the first six characters followed by a 'tilde' character "~" and a number, usually 1. More numbers (2, 3, and so on) were added after the 'tilde' if more than one file name with same first characters existed in the same directory.

The 16-bit version distributed with Windows 3.1x and Windows for Workgroups 3.1x installations had a Y2K issue due to lexicographic correlation between date representation and the ASCII character set; colons and semicolons replaced what should have been '2000'. Microsoft issued corrected binaries for all Windows 3.1x environments.[3]

Windows NT

File Manager was completely rewritten as a 32-bit application for Windows NT. This new version correctly handled long file names as well as NTFS file systems. It was included with Windows NT 3.1, 3.5, 3.51, and 4.0. It is possible to run File Manager on Windows 2000, Windows XP and Windows Server 2003 by extracting a copy of the WINFILE.EXE program file from a Windows NT 4.0 Installation or CD-ROM, or by downloading and extracting the file from Windows NT 4.0 Service Pack 6a (SP6a).

File Manager cannot run natively under Windows Vista, Windows Server 2008 or Windows 7 because these operating systems do not include COMMCTRL.DLL and other necessary resources (Dynamic Link Libraries). However, it is possible to create a modified version that will run on Vista, Server 2008 and 7 using manual instructions available here [4]. This functionality (plus related help and support files) is packaged into an MSI installer at this page [5] and into an IEAK installer at this page [6].

Undocumented Features

The 16-bit File manager cannot be configured as an orthodox file manager with two panels side-by-side using options in the menu, but with two directory windows open, if the user holds down the SHIFT key and selects WINDOW->TILE VERTICALLY the windows will form two identically-sized panels as shown.[7]

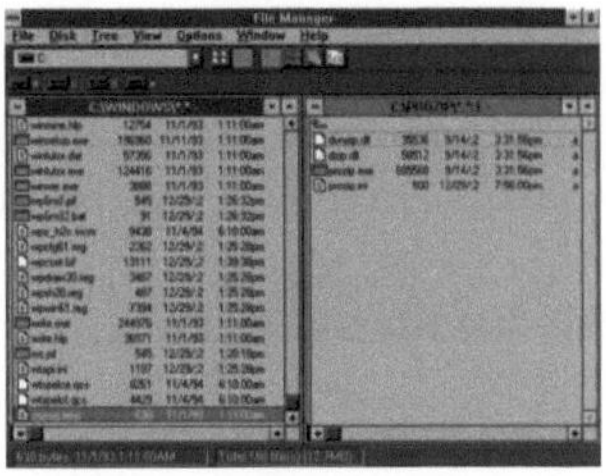

WINFILE.EXE configured as an orthodox file manager

See also

- DOS Shell

References

[1] http://www.mdgx.com/bin.htm#FM
[4] http://people.ee.ethz.ch/~davidsch/vistafm/
[5] http://www.brydon.net/winfile/
[6] http://www.mdgx.com/xptoy.htm#FMNT

External links

- Windows File Manager Revived (http://people.ee.ethz.ch/~davidsch/vistafm/) (guide for Windows Vista, should work also with Server 2008 and 7)
- Reviving Windows File Manager (http://www.brydon.net/winfile/) (WinFile.exe MSI installer for Windows Vista, works also with 7 and should work with Server 2008)
- File Manager (FM) & Add-ons (http://www.mdgx.com/xptoy.htm#FMNT) (WinFile.exe IEAK installer for all Windows NTx OSes)
- File Manager 16-bit and 32-bit Updates (http://www.mdgx.com/bin.htm#FM) (WinFile.exe IEAK installers for all Windows OSes and environments)

File_manager

A **file manager** or **file browser** is a computer program that provides a user interface to work with file systems. The most common operations performed on files or groups of files are: create, open, edit, view, print, play, rename, move, copy, delete, search/find, and modify file attributes, properties and file permissions. Files are typically displayed in a hierarchy. Some file managers contain features inspired by web browsers, including forward and back navigational buttons.

Some file managers provide network connectivity via protocols, such as FTP, NFS, SMB or WebDAV. This is achieved by allowing the user to browse for a file server (connecting and accessing the server's file system like a local file system) or by providing its own full client implementations for file server protocols.

Directory editors

A term that predates the usage of *file manager* is *directory editor*, which dates back at least to 1978.

A directory editor was written for EXEC 8 at the University of Maryland, and was available to other users at that time. The term was used by other developers, including Jay Lepreau, who wrote the **dired** program in 1980, [1] which ran on BSD. This was in turn inspired by an older program with the same name running on TOPS-20. **Dired** inspired other programs, including dired, the editor script (for emacs and similar editors), and **ded**. [2]

File-list file manager

File-list file managers are lesser known and older than orthodox file managers.

flist

One such file manager is **flist**, which was first used in 1981 on the Conversational Monitor System.[3] [4] This is a variant of **fulist**, which originated before late 1978, according to comments by its author, Theo Alkema.[5]

The **flist** program provided a list of files in the user's minidisk,[6] and allowed sorting by any file attribute. The file attributes could be passed to scripts or function-key definitions, making it simple to use flist as part of CMS EXEC, EXEC 2 or xedit scripts.

This program ran only on IBM VM/SP CMS, but was the inspiration for other programs, including **filelist**[7] [8] [9] (a script run via the Xedit editor), and programs running on other operating systems, including a program also called **flist**, which ran on OpenVMS,[10] and **fulist** (from the name of the corresponding internal IBM program),[11] which runs on Unix.[12]

Orthodox file managers

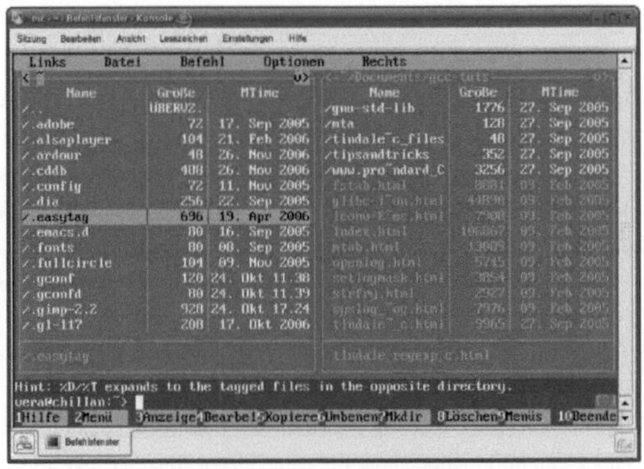

The well-known orthodox file manager Midnight Commander implements a Text-based user interface-based design

Orthodox file managers (OFM) or *command-based file managers* have three windows (two panels and one command line window). Orthodox file managers are one of the older families of file managers. Developers create applications that duplicate and extend the interface that was introduced by PathMinder and John Socha's famous Norton Commander for DOS. The concept is more than twenty five years old—PathMinder was released in 1984, and Norton Commander version 1.0 was released in 1986. Despite the age of this concept, file managers based on Norton Commander are actively developed, and dozens of implementations exist for DOS, Unix, and Microsoft Windows. Nikolai Bezroukov publishes his own set of criteria for an OFM standard (version 1.2 dated June 1997).[13]

Features

An orthodox file manager typically has three windows. Two of the windows are called panels and are positioned symmetrically at the top of the screen. The third is the command line, which is essentially a minimized command (shell) window that can be expanded to full screen. Only one of the panels is active at a given time. The active panel contains the "file cursor". Panels are resizable and can be hidden. Files in the active panel serve as the source of file operations performed by the manager. For example, files can be copied or moved from the active panel to the location represented in the passive panel. This scheme is most effective for systems in which the keyboard is the primary or sole input device. The active panel shows information about the current working directory and the files that it contains. The passive (inactive) panel shows the content of the same or another directory (the default target for file operations). Users may customize the display of columns that show relevant file information. The active panel and passive panel can be switched (often by pressing the tab key).

The following features describe the class of orthodox file managers.

- They present the user with a two-panel directory view with a command line below. Either panel may be selected to be active; the other becomes passive. The active panel becomes the working area for delete and rename operations, while the passive panel serves as a target for copy and move operations. Panels may be shrunk, exposing the terminal window hidden behind them. Normally, only the last line of the terminal window (the command line) is visible.
- They provide close integration with an underlying OS shell via command line, using the associated terminal window that permits viewing the results of executing shell commands entered on the command line (e.g., via Ctrl-O shortcut in Norton Commander).
- They provide the user with extensive keyboard shortcuts.
- The file manager frees the user from having to use the mouse.
- Users can create their own file associations and scripts that are invoked for certain file types and organize these scripts into a hierarchical tree (e.g., as a user script library or user menu).

- Users can extend the functionality of the manager via so-called *User menu* or *Start menu* and extensions menu. Norton Commander introduced the concept of user-defined file associations that is now used in all modern file managers.

Other common features include:

- Information on the "active" and "passive" panels may be used for constructing commands on the command line. Examples include current file, path to left panel, path to right panel, etc.
- They provide a built-in viewer for (at least) the most basic file types.
- They have a built-in editor. In many cases, the editor can extract certain elements of the panels into the text being edited.
- Many support virtual file systems (VFS) such as viewing compressed archives, or working with files via an FTP connection.
- They often have the word *commander* in the name, after Norton Commander.
- Path: shows the source/destination location of the directory in use
- Information about directory size, disk usage and disk name (usually at the bottom of the panels)
- Panel with information about file name, extension, date and time of creation, last modification, and permissions (attributes).
- Info panel with the number of files in directory, and the sum of the sizes of selected files.
- Tabbed interface (usually GUI file managers)
- Function keys: F1–F10 have all the same functions under all orthodox file managers. Examples: F5 always copies file(s) from the active to the inactive panel, while F6 moves the file.

Tabbed panels

The introduction of tabbed panels in some file managers (for example Total Commander) made it possible to manipulate more than one active and passive directory at a time.

Portability

Orthodox file managers[14] are among the most portable file managers. Examples are available on almost any platform, with both command-line interface and graphical user interface. This is unusual among command line managers in that something purporting to be a standard for the interface is published. They are also actively supported by developers. This makes it possible to do the same work on different platforms without much relearning of the interface.

Dual-pane managers

Sometimes they are called dual-pane managers, a term that is typically used for programs such as the Windows File Explorer (see below). But they have three panes including a command line pane below (or hidden behind) two symmetric panes. Furthermore, most of these programs allow using just one of the two larger panes with the second hidden.

Examples of Orthodox file managers

Notable ones include:

- Altap Salamander
- Demos Commander
- Directory Opus
- Dos Navigator
- emelFM2
- FAR Manager

- File Commander
- GNOME Commander
- Krusader
- Midnight Commander
- muCommander
- Norton Commander
- PathMinder
- SE-Explorer
- Total Commander
- Volkov Commander
- WinSCP
- XTree
- ZTreeWin

Navigational file manager

A navigational file manager, also called an "Explorer" type manager, is a newer type of file manager which became prominent because of its integration in Microsoft Windows. The Windows Explorer is a classic representative of the type, using a "navigational" metaphor to represent filesystem locations. Since the advent of GUIs, it has become the dominant type of file manager for desktop computers, due to its use in the Microsoft Windows operating system.

Typically, it has two panes, with the filesystem tree in the left pane and the contents of the current directory in the right pane. For Mac OS X, one view in the Finder is an example of a navigational file manager.

Concepts

- The window displays the location currently being viewed.
- The location being viewed (the current directory) can be changed by the user by opening directories, pressing a *back button*, typing a location, or using the additional pane with the navigation tree representing all or part of the filesystem.
- Icons represent files, programs, and directories.

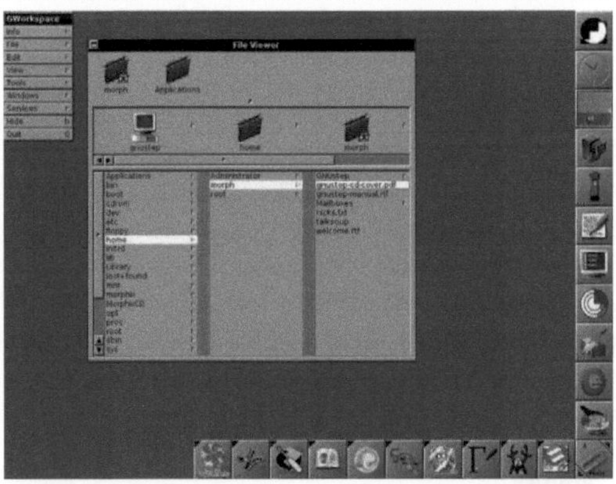

The Miller Column browser from GNUstep is a type of Navigational file manager.

The interface in a navigational file manager often resembles a web browser, complete with *back* and *forward* buttons, and often *reload* buttons. Most also contain an address bar into which the file or directory path (or URI) can be typed.

Most navigational file managers have two panes, the left pane being a tree view of the filesystem. This means that unlike orthodox file managers, the two panes are asymmetrical in their content and use.

Selecting a directory in the Navigation pane on the left designates it as the current directory, displaying its contents in the Contents pane on the right. However, expanding (+) or collapsing (-) a portion of the tree without selecting a directory will not alter the contents of the right pane. The exception to this behavior applies when collapsing a parent of the current directory, in which case the selection is refocused on the collapsed parent directory, thus altering the list in the Contents pane.

The process of moving from one location to another need not open a new window. Several instances of the file manager can be opened simultaneously and communicate with each other via drag-and-drop and clipboard operations, so it is possible to view several directories simultaneously and perform cut-and paste operations between instances.

File operations are based on drag-and-drop and editor metaphors: users can select and copy files or directories onto the clipboard and then paste them in a different place in the filesystem or even in a different instance of the file manager.

Examples

Notable examples of navigational file managers include:

- File Manager
- Windows Explorer
- Mac OS X Finder
- Directory Opus
- XTree / ZTreeWin
- XYplorer

Spatial file manager

Spatial file managers use a spatial metaphor to represent files and directories as if they were actual physical objects. A spatial file manager imitates the way people interact with physical objects.

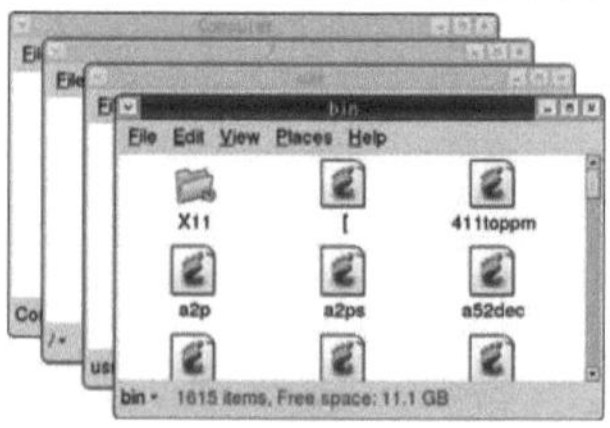

The Nautilus file manager has a spatial mode. Each of these windows displays an open directory.

Some ideas behind the concept of a spatial file manager are:

1. A single window represents each opened directory
2. Each window is unambiguously and irrevocably tied to a particular directory.
3. Stability: files, directories, and windows go where the user moves them, stay where the user puts them ("preserve their spatial state"), and retain all their other "physical" characteristics (such as size, shape, color and location).
4. The same item can only be viewed in one window at a time.

As in navigational file managers, when a directory is opened, the icon representing the directory changes—perhaps from an image showing a closed drawer to an opened one, perhaps the directory's icon turns into a silhouette filled with a pattern—and a new window is opened to represent that directory.

Examples

Examples of file managers that use a spatial metaphor to some extent include:

- Apple's Finder 5 to 9 (In the Mac OS X Finder the spatial metaphor can still be used in general by enabling Finder's preference "Always open folders in a new window" [15] or on demand by holding the command key () while double-clicking a folder [16])
- Konqueror has the option to turn into spatial mode
- RISC OS Filer
- Amiga's Workbench
- GNOME's Nautilus from version 2.6 (default until 2.29)
- MATE's Caja (though the default mode is navigational)
- BeOS's Tracker

- Haiku's Tracker
- OS/2's Workplace Shell
- Digital Research's GEM (implemented in Atari TOS and as a somewhat reduced version for PCs)
- ZDESKTOP and FILEMAGE [17] Zoomable File-System Viewers (spatial view of hierarchical data)
- ROX-Filer file manager (ROX Desktop)
- E17 file manager

Dysfunctional spatial file managers:

- Windows Explorer in Windows 95 was set as a spatial file manager model by default; because it also worked as a navigational file manager, directories could be opened in multiple windows, which made it fail all the above criteria. Later versions gradually abandoned the spatial model.
- Apple's Finder in Mac OS X was designed with a similar integration of spatial and navigational modes, meaning that the spatial mode did not actually work.[18]

3D file managers

Some projects have attempted to implement a three-dimensional method of displaying files and directory structures. Three-dimensional file browsing has not yet become popular; the exact implementation tends to differ between projects, and there are no common standards to follow.

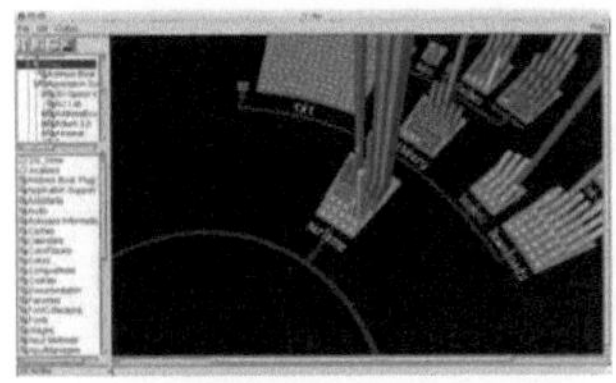

File System Visualizer, one example of a 3D file manager.

Examples

Examples of three-dimensional file managers include:

- fsn, for Silicon Graphics' IRIX systems, notably featured prominently in one scene from the film Jurassic Park, as a representation of Unix systems.
- File System Visualizer, or fsv, an open-source clone of fsn for modern Unix-like systems.
- BumpTop, a file manager using a three-dimensional representation of a desktop with realistic physics, intended for use with a stylus and touchscreen.
- Real Desktop,[19] a desktop replacement with similarities to BumpTop.
- Cubix 3D Filer [20] is a Windows file manager which organizes files according to different attributes.

Web-based file managers

Web-based file managers are typically scripts written in either PHP, Ajax, Perl, ASP or another server-side language. When installed on a local server or on a remote server, they allow files and directories located there to be managed and edited, using a web browser, without the need for FTP Access.

More advanced, and usually commercially-distributed, web-based file management scripts allow the administrator of the file manager to configure secure, individual user accounts, each with individual account permissions. Authorized users have access to documents stored on the server or in their individual user directories anytime, from anywhere, via a web browser.

A web-based file manager can serve as an organization's digital repository. For example, documents, digital media, publishing layouts, and presentations can be stored, managed, and shared between customers, suppliers, and remote workers, or just internally.

Web-based file managers are becoming increasingly popular due to the rise in popularity of dynamic web content management systems (CMS) and the need for non-technical website moderators to manage media on their websites powered by these platforms.

Examples

- net2ftp – a popular, PHP and JavaScript-based FTP client

Popular culture

A 3D file manager is featured in *Jurassic Park*, during a scene where Lex desperately tries to find an executable file, while a Velociraptor tries to force its way into the command center. The 3D file manager, fsn (mentioned above), was built on top of a Unix system (Silicon Graphics, Inc's IRIX).

See also

- Batch renaming
- Comparison of file managers
- Computer file management
- Desktop metaphor
- Miller Columns
- Spatial navigation

References

[17] http://andrea.l.vitali.googlepages.com/zdesktop
[19] product homepage
[20] The Cubix Project Homepage

External links

- File manager (http://www.dmoz.org/Computers/Software/File_Management/File_Managers/) at the Open Directory Project
- The Orthodox File Manager (OFM) Paradigm: The History of Development of Norton Commander (http://www.softpanorama.org/OFM/Paradigm/Ch03/norton_commander.shtml), by Dr. Nikolai Bezroukov, Softpanorama.org, 2009. Retrieved 2010-12-26.
- Less is More: A rich functionality behind Spartan interface of Orthodox File Managers (http://www.softpanorama.org/Articles/introduction_to_orthodox_file_managers.shtml), by Dr. Nikolai Bezroukov, Softpanorama.org, 2012. Retrieved 2012-12-15.
- About the Finder (http://arstechnica.com/articles/paedia/finder.ars), by John Siracusa, Ars Technica, 2003. Retrieved 2010-12-26.
- The Spatial Way (http://www.bytebot.net/geekdocs/spatial-nautilus.html), by Colin Charles, 2004. Retrieved 2010-12-26.
- dired - directory editor (http://web.archive.org/web/20080403063510/http://freeware.sgi.com/fw-5.3/fw_dired/dired.html). Archived from the original (http://freeware.sgi.com/fw-5.3/fw_dired/dired.html) on 2008-04-03. Retrieved 2010-12-26.
- flist (http://www.cs.uregina.ca/Dept/manuals/Manuals/4System/4_3cms/4_3CMS.html), section 4.3.2.2.3 in Introduction to IBM/CMS, Users' Manual, Department of Computer Science, University of Regina, Saskatchewan, Canada. Retrieved 2010-12-26.

Microsoft_Windows

Microsoft Windows is a series of graphical interface operating systems developed, marketed, and sold by Microsoft. Microsoft introduced an operating environment named *Windows* on November 20, 1985 as a graphical operating system shell for MS-DOS in response to the growing interest in graphical user interfaces (GUI).[] Microsoft Windows came to dominate the world's personal computer market with over 90% market share, overtaking Mac OS, which had been introduced in 1984.

As of April 2013, the most recent versions of Windows for personal computers, mobile devices, server computers and embedded devices are respectively Windows 8, Windows Phone 8, Windows Server 2012 and Windows Embedded 8.

Version history

The term *Windows* collectively describes any or all of several generations of Microsoft operating system products. These products are generally categorized as follows:

Early versions

The history of Windows dates back to September 1981, when Chase Bishop, a computer scientist, designed the first model of an electronic device and project "Interface Manager" was started. It was announced in November 1983 (after the Apple Lisa, but before the Macintosh) under the name "Windows", but Windows 1.0 was not released until November 1985.[1] Windows 1.0 achieved little popularity and was to compete with Apple's own operating system. Windows 1.0 is not a complete operating system; rather, it extends MS-DOS. The shell of Windows 1.0 is a program known as the MS-DOS Executive. Included components include Calculator, Calendar, Cardfile, Clipboard viewer, Clock, Control Panel, Notepad, Paint, Reversi, Terminal and Write. Windows 1.0 does not allow overlapping windows. Instead all windows are tiled. Only modal dialog boxes may appear over other windows.

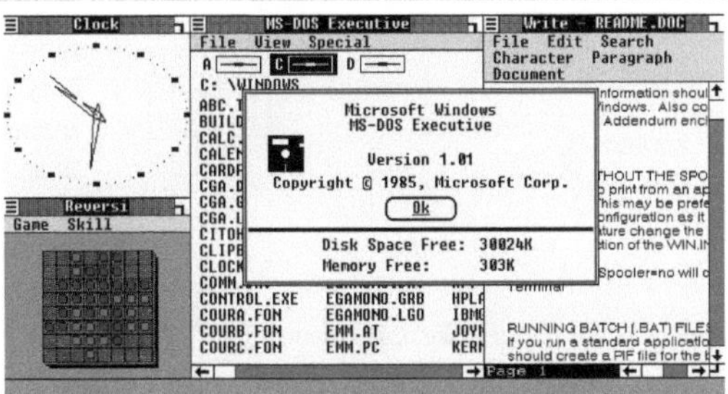

Windows 1.0, the first version, released in 1985

Windows 2.0 was released in December 1987 and was more popular than its predecessor. It features several improvements to the user interface and memory management. Windows 2.03 changed the OS from tiled windows to overlapping windows. The result of this change led to Apple Computer filing a suit against Microsoft alleging infringement on Apple's copyrights.[2] [3] Windows 2.0 also introduced more sophisticated keyboard shortcuts and could make use of expanded memory.

Windows 2.1 was released in two different versions: Windows/286 and Windows/386. Windows/386 uses the virtual 8086 mode of Intel 80386 to multitask several DOS programs and the paged memory model to emulate expanded memory using available extended memory. Windows/286, in spite of its name, runs on both Intel 8086 and Intel 80286. It runs in real mode but can make use of the high memory area.

In addition to full Windows-packages, there were runtime-only versions that shipped with early Windows software from third parties and made it possible to run their Windows software on MS-DOS and without the full Windows feature set.

The early versions of Windows are often thought of as graphical shells, mostly because they ran on top of MS-DOS and use it for file system services.[] However, even the earliest Windows versions already assumed many typical operating system functions; notably, having their own executable file format and providing their own device drivers (timer, graphics, printer, mouse, keyboard and sound). Unlike MS-DOS, Windows allowed users to execute multiple graphical applications at the same time, through cooperative multitasking. Windows implemented an elaborate, segment-based, software virtual memory scheme, which allows it to run applications larger than available memory: code segments and resources are swapped in and thrown away when memory became scarce; data segments moved in memory when a given application had relinquished processor control.

Windows 3.0 and 3.1

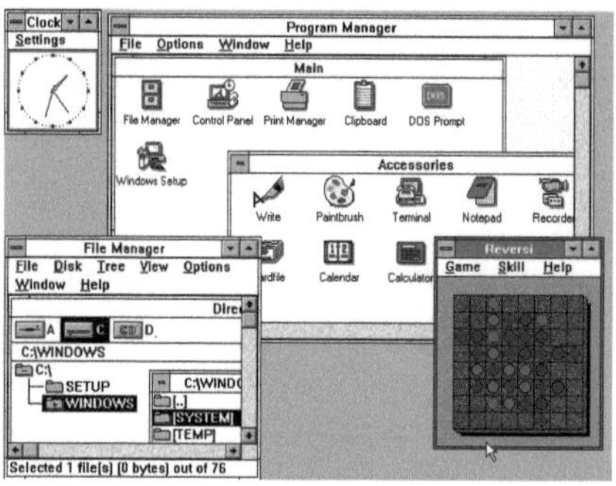

Windows 3.0, released in 1990

Windows 3.0, released in 1990, improved the design, mostly because of virtual memory and loadable virtual device drivers (VxDs) that allow Windows to share arbitrary devices between multi-tasked DOS applications. Windows 3.0 applications can run in protected mode, which gives them access to several megabytes of memory without the obligation to participate in the software virtual memory scheme. They run inside the same address space, where the segmented memory provides a degree of protection. Windows 3.0 also featured improvements to the user interface. Microsoft rewrote critical operations from C into assembly. Windows 3.0 is the first Microsoft Windows version to achieve broad commercial success, selling 2 million copies in the first six months.[4] [5]

Windows 3.1, made generally available on March 1, 1992, featured a facelift. In August 1993, Windows for Workgroups, a special version with integrated peer-to-peer networking features and a version number of 3.11, was released. It was sold along Windows 3.1. Support for Windows 3.1 ended on December 31, 2001.[6]

Windows 9x

Windows 95, released in August 1995, introduced the taskbar and Start menu to the operating system.

The next major consumer-oriented release of Windows, Windows 95, was released on August 24, 1995. While still remaining MS-DOS-based, Windows 95 introduced support for native 32-bit applications, plug and play hardware, preemptive multitasking, long file names of up to 255 characters, and provided increased stability over its predecessors. Windows 95 also introduced a redesigned, object oriented user interface, replacing the previous Program Manager with the Start menu, taskbar, and Windows Explorer shell. Windows 95 was a major commercial success for Microsoft; Ina Fried of CNET remarked that "by the time Windows 95 was finally ushered off the market in 2001, it had become a fixture on computer desktops around the world."[7] Microsoft published four OEM Service Releases (OSR) of Windows 95, each of which was roughly equivalent to a service pack. The first OSR of Windows 95 was also the first version of Windows to be bundled with Microsoft's web browser, Internet Explorer.[] Mainstream support for

Windows 95 ended on December 31, 2000 and extended support for Windows 95 ended on December 31, 2001.[]

Windows 95 was followed up with the release of Windows 98 on June 25, 1998, which introduced the Windows Driver Model, support for USB composite devices, support for ACPI, hibernation, and support for multi-monitor configurations. Windows 98 also included integration with Internet Explorer 4 through Active Desktop and other aspects of the Windows Desktop Update (a series of enhancements to the Explorer shell which were also made available for Windows 95). In May 1999, Microsoft released Windows 98 Second Edition, an updated version of Windows 98. Windows 98 SE added Internet Explorer 5.0 and Windows Media Player 6.2 amongst other upgrades. Mainstream support for Windows 98 ended on June 30, 2002 and extended support for Windows 98 ended on July 11, 2006.[]

On September 14, 2000, Microsoft released Windows ME (Millennium Edition), the last DOS-based version of Windows. Windows ME incorporated visual interface enhancements from its Windows NT-based counterpart Windows 2000, had faster boot times than previous versions (which however, required the removal of the ability to access a real mode DOS environment, removing compatibility with some older programs),[] expanded multimedia functionality (including Windows Media Player 7, Windows Movie Maker, and the Windows Image Acquisition framework for retrieving images from scanners and digital cameras), additional system utilities such as System File Protection and System Restore, and updated home networking tools.[] However, Windows ME was faced with criticism for its speed and instability, along with hardware compatibility issues and its removal of real mode DOS support. PC World considered Windows ME to be one of the worst operating systems Microsoft had ever released, and the 4th worst tech product of all time.[]

Windows NT family

Windows logo (1992–2001), introduced in Windows 3.1

Windows logo (2001–2006), introduced in Windows XP

Windows logo used as of October 2012, introduced in Windows Server 2012

In November 1988, a new development team within Microsoft (which included former Digital Equipment Corporation developers Dave Cutler and Mark Lucovsky) began work on a revamped version of IBM and Microsoft's OS/2 operating system known as "NT OS/2". NT OS/2 was intended to be a secure, multi-user operating system with POSIX compatibility and a modular, portable kernel with preemptive multitasking and support for multiple processor architectures. However, following the successful release of Windows 3.0, the NT development team decided to rework the project to use an extended 32-bit port of the Windows API known as Win32 instead of those of OS/2. Win32 maintained a similar structure to the Windows APIs (allowing existing Windows applications to easily be ported to the platform), but also supported the capabilities of the existing NT kernel. Following its approval by Microsoft's staff, development continued on what was now Windows NT, the first 32-bit version of Windows. However, IBM objected to the changes, and ultimately continued OS/2 development on its own.[] []

The first release of the resulting operating system, Windows NT 3.1 (named to associate it with Windows 3.1) was released in July 1993 with versions for desktop workstations and servers. Windows NT 3.5 was released in September 1994, focusing on performance improvements and support for Novell's NetWare, and was followed up by Windows NT 3.51 in May 1995, which included additional improvements and support for the PowerPC architecture. Windows NT 4.0 was released in June 1996, introducing the redesigned interface of Windows 95 to the NT series. On February 17, 2000, Microsoft released Windows 2000, a successor to NT 4.0. The Windows NT name was dropped at this point in order to put a greater focus on the Windows brand.[]

Windows XP

The next major version of Windows, Windows XP, was released on October 25, 2001. The introduction of Windows XP aimed to unify the consumer-oriented Windows 9x series with the architecture introduced by Windows NT, a change which Microsoft promised would provide better performance over its DOS-based predecessors. Windows XP would also introduce a redesigned user interface (including an updated Start menu and a "task-oriented" Windows Explorer), streamlined multimedia and networking features, Internet Explorer 6, integration with Microsoft's .NET Passport services, modes to help provide compatibility with software designed for previous versions of Windows, and Remote Assistance functionality.[]

At retail, Windows XP was now marketed in two main editions: the "Home" edition was targeted towards consumers, while the "Professional" edition was targeted towards business environments and power users, and included additional security and networking features. Home and Professional were later accompanied by the "Media Center" edition (designed for home theater PCs, with an emphasis on support for DVD playback, TV tuner cards, DVR functionality, and remote controls), and the "Tablet PC" edition (designed for mobile devices meeting its specifications for a tablet computer, with support for stylus pen input and additional pen-enabled applications).[8] [9] [10] Mainstream support for Windows XP ended on April 14, 2009. Extended support will continue until April 8, 2014.[11]

After Windows 2000, Microsoft also changed its release schedules for server operating systems; the server counterpart of Windows XP, Windows Server 2003, was released in April 2003.[] It was followed in December 2005 by Windows Server 2003 R2.

Windows Vista, 7 and 8

After a lengthy development process, Windows Vista was released on November 30, 2006 for volume licensing and January 30, 2007 for consumers. It contained a number of new features, from a redesigned shell and user interface to significant technical changes, with a particular focus on security features. It was available in a number of different editions, and has been subject to some criticism. Vista's server counterpart, Windows Server 2008 was released in early 2008.

On July 22, 2009, Windows 7 and Windows Server 2008 R2 were released as RTM (release to manufacturing) while the former was released to the public 3 months later on October 22, 2009. Unlike its predecessor, Windows Vista, which introduced a large number of new features, Windows 7 was intended to be a more focused, incremental upgrade to the Windows line, with the goal of being compatible with applications and hardware with which Windows Vista was already compatible.[12] Windows 7 has multi-touch support, a redesigned Windows shell with an updated taskbar, a home networking system called HomeGroup,[13] and performance improvements.

Windows 8, the successor to Windows 7, was released generally on October 28, 2012. A number of significant changes were made on Windows 8, including the introduction of a user interface based around Microsoft's Metro design language with optimizations for touch-based devices such as tablets and all-in-one PCs. These changes include the Start screen, which uses large tiles that are more convenient for touch interactions and allow for the display of continually updated information, and a new class of apps which are designed primarily for use on touch-based devices. Other changes include increased integration with cloud services and other online platforms (such as social networks and Microsoft's own SkyDrive and Xbox Live services), the Windows Store service for software distribution, and a new variant known as Windows RT for use on devices that utilize the ARM architecture.[] [] [] [] [] []

Multilingual support: IMEs and LIPs

There are three main issues involved in making English-language Windows multilingual: (1) some languages require an Input Method Editor (IME) to enter text, (2) many users will want application menus (such as MS Office menus) to display in their own language, and they may also want to use a keyboard that matches the normal keyboard layout and marking for their own language, and (3) some users will want Windows menus and messages to display in their own language, i.e. they will want to switch from an English Windows environment to another language.

For languages like Italian, Spanish, French and German, (2) alone may suffice. For languages like Chinese, Japanese, and Korean (CJK), an IME (1) is also required. This is bundled with the corresponding language version of Windows, but is also available as a separate download for English Windows, as described below; (1) and (2) can be essentially free (apart from the custom keyboard). For some languages, (3), multilingual support for Windows, is a free download for Windows XP and later—but it requires Windows 7 Ultimate or better for languages such as Chinese, Japanese, and Korean.

(1) After releasing Chinese, Japanese, and Korean bundles of Office 2010 and IME 2010, Microsoft made IME 2010 available as a free upgrade for users of the earlier IME versions of Windows. Microsoft later made these Chinese, Japanese, and Korean IME versions available free to users of Windows XP and later, including English Windows XP (but now says that users should own some version of MS Office).[14] Each IME package enables the entering of text in the corresponding language; necessary fonts may be bundled with it (or supplementary fonts offered with the corresponding version of Office).

(2) Microsoft now also offers Language Interface Packs (LIPs) for MS Office. Some LIPs are free;[15] some "Language Packs" (such as the CJK ones) are sold separately and may include spelling and grammar checking tools.[16] (Recent application software from some companies may support two or more popular languages).

(3) Microsoft now also offers Language Interface Packs (LIPs) that allow users to view Windows menus, dialog boxes, and other user interface items in their preferred language. These are free; most are for English Windows (XP and later)—however, Chinese, Japanese, and Korean LIP downloads require Windows 7 Ultimate or Enterprise.[17] These LIPs include IMEs where applicable.

Platform support

Windows NT included support for several different platforms before the x86-based personal computer became dominant in the professional world. Windows NT 4.0 and its predecessors supported PowerPC, DEC Alpha and MIPS R4000. (Although some these platforms implement 64-bit computing, the operating system treated them as 32-bit.) However, Windows 2000, the successor of Windows NT 4.0, dropped support for all platforms except the third generation x86 (known as IA-32) or newer in 32-bit mode. The client line of Window NT family still runs on IA-32, although the Windows Server line has ceased supporting this platform with the release of Windows Server 2008 R2.

With the introduction of the Intel Itanium architecture (IA-64), Microsoft released new versions of Windows to support it. Itanium versions of Windows XP and Windows Server 2003 were released at the same time as their mainstream x86 counterparts. Windows XP 64-Bit Edition, released in 2005, is the last Windows client operating systems to support Itanium. Windows Server line continued to support this platform until Windows Server 2012; Windows Server 2008 R2 is the last Windows operating system to support Itanium architecture.

On April 25, 2005, Microsoft released Windows XP Professional x64 Edition and Windows Server 2003 x64 Editions to support the x86-64 (or simply x64), the eighth generation of x86 architecture. Windows Vista was the first client version of Windows NT to be released simultaneously in IA-32 and x64 editions. x64 is still supported.

An edition of Windows 8 known as Windows RT was specifically created for computers with ARM architecture.

Windows CE

Windows CE (officially known as *Windows Embedded Compact*), is an edition of Windows that runs on minimalistic computers, like satellite navigation systems and some mobile phones. Windows Embedded Compact is based on its own dedicated kernel, dubbed Windows CE kernel. Microsoft licenses Windows CE to OEMs and device makers. The OEMs and device makers can modify and create their own user interfaces and experiences, while Windows CE provides the technical foundation to do so.

The latest current version of Windows CE, Windows Embedded Compact 7, displaying a concept media player UI.

Windows CE was used in the Dreamcast along with Sega's own proprietary OS for the console. Windows CE was the core from which Windows Mobile was derived. Its successor, Windows Phone 7, was based on components from both Windows CE 6.0 R3 and Windows CE 7.0. Windows Phone 8 however, is based on the same NT-kernel as Windows 8.

Windows Embedded Compact is not to be confused with Windows XP Embedded or Windows NT 4.0 Embedded, modular editions of Windows based on Windows NT kernel.

Timeline of releases

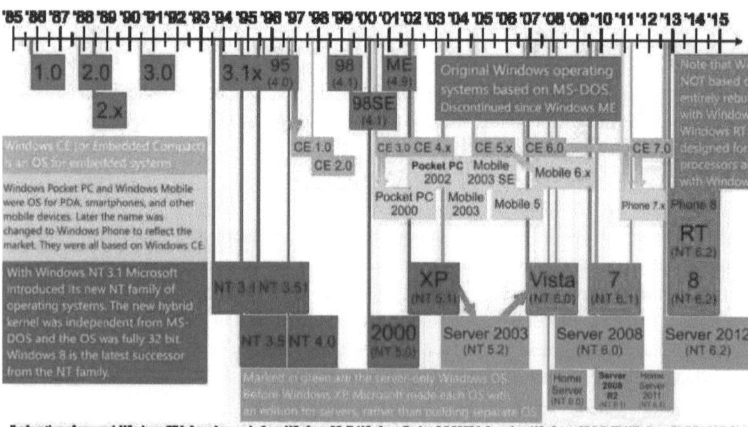

Explanation of arrows: I. Windows CE is based on code from Windows 95. II. Windows Pocket PC 2000 is based on Windows CE 3.0. III. Windows Mobile 6.x is based on Windows CE 5.x, rather than CE 6.0. IV. Windows Phone 7 is based on code from both Windows CE 6.0 and CE 7.0. V. Windows Vista was built on code from Windows Server 2003, rather than Windows XP.

The Windows family tree

Usage share

Source	Net Market Share[18]	Global Stats[19]	W3Counter[20]
Date	July 2013	July 2013	July 2013
All versions	91.36%	84.78%	72.19%
Windows 7	44.49%	52.48%	42.76%
Windows XP	37.19%	20.45%	20.08%
Windows 8	5.4%	6.61%	4.86%
Windows Vista	4.24%	5.24%	4.49%
Windows 2000	0.04%	—	—

Security

Consumer versions of Windows were originally designed for ease-of-use on a single-user PC without a network connection, and did not have security features built in from the outset.[21] However, Windows NT and its successors are designed for security (including on a network) and multi-user PCs, but were not initially designed with Internet security in mind as much, since, when it was first developed in the early 1990s, Internet use was less prevalent.[22]

These design issues combined with programming errors (e.g. buffer overflows) and the popularity of Windows means that it is a frequent target of computer worm and virus writers. In June 2005, Bruce Schneier's *Counterpane Internet Security* reported that it had seen over 1,000 new viruses and worms in the previous six months.[23] In 2005, Kaspersky Lab found around 11,000 malicious programs—viruses, Trojans, back-doors, and exploits written for Windows.[]

Microsoft releases security patches through its Windows Update service approximately once a month (usually the second Tuesday of the month), although critical updates are made available at shorter intervals when necessary.[24] In versions of Windows after and including Windows 2000 SP3 and Windows XP, updates can be automatically downloaded and installed if the user selects to do so. As a result, Service Pack 2 for Windows XP, as well as Service Pack 1 for Windows Server 2003, were installed by users more quickly than it otherwise might have been.[25]

While the Windows 9x series offered the option of having profiles for multiple users, they had no concept of access privileges, and did not allow concurrent access; and so were not true multi-user operating systems. In addition, they implemented only partial memory protection. They were accordingly widely criticised for lack of security.

The Windows NT series of operating systems, by contrast, are true multi-user, and implement absolute memory protection. However, a lot of the advantages of being a true multi-user operating system were nullified by the fact that, prior to Windows Vista, the first user account created during the setup process was an administrator account, which was also the default for new accounts. Though Windows XP did have limited accounts, the majority of home users did not change to an account type with fewer rights – partially due to the number of programs which unnecessarily required administrator rights – and so most home users ran as administrator all the time.

Windows Vista changes this[26] by introducing a privilege elevation system called User Account Control. When logging in as a standard user, a logon session is created and a token containing only the most basic privileges is assigned. In this way, the new logon session is incapable of making changes that would affect the entire system. When logging in as a user in the Administrators group, two separate tokens are assigned. The first token contains all privileges typically awarded to an administrator, and the second is a restricted token similar to what a standard user would receive. User applications, including the Windows Shell, are then started with the restricted token, resulting in a reduced privilege environment even under an Administrator account. When an application requests higher privileges or "Run as administrator" is clicked, UAC will prompt for confirmation and, if consent is given (including administrator credentials if the account requesting the elevation is not a member of the administrators group), start the process using the unrestricted token.[]

File permissions

All Windows versions from Windows NT 3 have been based on a file system permission system referred to as AGLP (Accounts, Global, Local, Permissions) AGDLP which in essence where file permissions are applied to the file/folder in the form of a 'local group' which then has other 'global groups' as members. These global groups then hold other groups or users depending on different Windows versions used. This system varies from other vendor products such as Linux and NetWare due to the 'static' allocation of permission being applied directory to the file or folder. However using this process of AGLP/AGDLP/AGUDLP allows a small number of static permissions to be applied and allows for easy changes to the account groups without reapplying the file permissions on the files and folders.

Windows Defender

On January 6, 2005, Microsoft released a Beta version of Microsoft AntiSpyware, based upon the previously released Giant AntiSpyware. On February 14, 2006, Microsoft AntiSpyware became Windows Defender with the release of Beta 2. Windows Defender is a freeware program designed to protect against spyware and other unwanted software. Windows XP and Windows Server 2003 users who have genuine copies of Microsoft Windows can freely download the program from Microsoft's web site, and Windows Defender ships as part of Windows Vista and 7.[] In Windows 8, Windows Defender and Microsoft Security Essentials have been combined into a single program, named Windows Defender. It is based on Microsoft Security Essentials, borrowing its features and user interface. Although it is enabled by default, it can be turned off to use another anti-virus solution.[] Windows Malicious Software Removal Tool and the optional Microsoft Safety Scanner are two other free security products offered by Microsoft.

Third-party analysis

In an article based on a report by Symantec,[] internetnews.com has described Microsoft Windows as having the "fewest number of patches and the shortest average patch development time of the five operating systems it monitored in the last six months of 2006."[27]

A study conducted by Kevin Mitnick and marketing communications firm Avantgarde in 2004 found that an unprotected and unpatched Windows XP system with Service Pack 1 lasted only 4 minutes on the Internet before it was compromised, and an unprotected and also unpatched Windows Server 2003 system was compromised after being connected to the internet for 8 hours.[] This study does not apply to Windows XP systems running the Service Pack 2 update (released in late 2004), which vastly improved the security of Windows XP. The computer that was running Windows XP Service Pack 2 was not compromised. The AOL National Cyber Security Alliance Online Safety Study of October 2004 determined that 80% of Windows users were infected by at least one spyware/adware product. Much documentation is available describing how to increase the security of Microsoft Windows products. Typical suggestions include deploying Microsoft Windows behind a hardware or software firewall, running anti-virus and anti-spyware software, and installing patches as they become available through Windows Update.[28]

Emulation software

Emulation allows the use of some Windows applications without using Microsoft Windows. These include:

- **Wine** – a free and open-source implementation of the Windows API, allowing one to run many Windows applications on x86-based platforms, including UNIX, Linux and OS X. Wine developers refer to it as a "compatibility layer"[29] and use Windows-style APIs to emulate Windows environment.
  - **CrossOver** – a Wine package with licensed fonts. Its developers are regular contributors to Wine, and focus on Wine running officially supported applications.
  - **Cedega** – proprietary fork of Wine by TransGaming Technologies, designed specifically for running Microsoft Windows games on Linux. A version of Cedega known as Cider allows Windows games to run on OS X. Since Wine was licensed under the LGPL, Cedega has been unable to port the improvements made to Wine to their proprietary codebase. Cedega ceased its service in February 2011.
  - **Darwine** – a port of Wine for OS X and Darwin. Operates by running Wine on QEMU.
- **ReactOS** – an open-source OS intended to run the same software as Windows, originally designed to simulate Windows NT 4.0, now aiming at Windows 7 compatibility. It has been in the development stage since 1996.
- **Linspire** – formerly LindowsOS, a commercial Linux distribution initially created with the goal of running major Windows software. Changed its name to Linspire after Microsoft v. Lindows. Discontinued in favor of Xandros Desktop.
- **Freedows OS** – an open-source attempt at creating a Windows clone for x86 platforms, intended to be released under the GNU General Public License. Started in 1996 by Reece K. Sellin, the project was never completed, getting only to the stage of design discussions which featured a number of novel concepts until it was suspended in 2002.[30] [31] [32]

See also

- Architecture of the Windows NT operating system line
- Criticism of Microsoft Windows
- Comparison of operating systems
- Comparison of Windows versions
- List of Microsoft Windows components
- List of operating systems
- Wintel
- Win-OS/2

References

[1] A history of Windows (http://windows.microsoft.com/en-in/windows/history) (at microsoft.com)

[14] Download IME 2010 (http://www.microsoft.com/en-US/download/details.aspx?id=30390) Microsoft.

[15] MS Office Language Interface Packs (http://office.microsoft.com/en-us/downloads/office-language-interface-pack-lip-downloads-HA001113350.aspx) Microsoft.

[16] MS Office Language Packs (http://www.microsoftstore.com/store/msstore/list/parentCategoryID.37286600/categoryID.54807000) Microsoft.

[17] Download Windows Language Interface Packs (http://windows.microsoft.com/en-US/windows/language-packs) Microsoft.

[21] Multi-user memory protection was not introduced until Windows NT and XP, and a computer's default user was an administrator until Windows Vista. Source: UACBlog (http://blogs.msdn.com/uac/).

[26] Microsoft describes in detail the steps taken to combat this in a TechNet bulletin. Source: Windows Vista Security and Data Protection Improvements (http://technet.microsoft.com/en-us/windowsvista/aa905073.aspx).

[29] Wine (http://www.winehq.org/)

External links

- Microsoft Developer Network (http://www.msdn.com/)
- Windows Client Developer Resources (http://msdn.microsoft.com/en-au/windows/default.aspx?WT.mc_id=soc-c-au-loc--2010oct)
- Microsoft Windows History Timeline (http://www.microsoft.com/Windows/WinHistoryIntro.mspx)
- Pearson Education, InformIT (http://www.informit.com/articles/article.aspx?p=1358665&rll=1) – History of Microsoft Windows
- Microsoft Windows 7 for Government (http://www.microsoft.com/industry/government/products/windows7/default.aspx)

MS-DOS

MS-DOS (/ˌɛmɛsˈdɒs/ ***EM-es-DOSS***; short for Microsoft Disk Operating System) is an operating system for x86-based personal computers. It was the most commonly used member of the DOS family of operating systems, and was the main operating system for IBM PC compatible personal computers during the 1980s to the mid-1990s, until it was gradually superseded by operating systems offering a graphical user interface (GUI), in particular by various generations of the Microsoft Windows operating system.

MS-DOS grew out of a request placed by IBM in 1981 for an operating system to use in its IBM PC range of personal computers. Microsoft quickly bought the rights to QDOS (Quick and Dirty Operating System), also known as 86-DOS,[1] from Seattle Computer Products, and began work on modifying it to meet IBM's specification, who licensed and released it as PC DOS 1.0 to be used with their PCs in August 1981. Although MS-DOS and PC DOS were initially developed in parallel by Microsoft and IBM, years later the two products eventually went their separate ways.

During its life, several competing products were released for the x86 platform,[] and MS-DOS itself would go through eight versions, until development ceased in 2000. Ultimately it was the key product in Microsoft's growth from a programming languages company to a diverse software development firm, providing the company with essential revenue and marketing resources. It was also the underlying basic operating system on which early versions of Windows ran as a GUI. It is a flexible operating system, and consumes only 8 MB of installation space.

History

MS-DOS was a renamed form of 86-DOS – informally known as the Quick-and-Dirty Operating System or Q-DOS[] – owned by Seattle Computer Products, written by Tim Paterson.[] Microsoft needed an operating system for the then-new Intel 8086 but it had none available, so it bought 86-DOS 1.10 for $75,000.[] Development of 86-DOS took only six weeks, as it was basically a clone of Digital Research's CP/M (for 8080/Z80 processors), ported to run on 8086 processors and with two notable differences compared to CP/M, an improved disk sector buffering logic and the introduction of FAT12 instead of the CP/M filesystem. This first version was shipped in August 1980[] until Microsoft hired Tim Paterson in May 1981 and acquired 86-DOS 1.10 in July of the same year. Microsoft kept the version number, but renamed it MS-DOS. They also licensed MS-DOS 1.10/1.14 to IBM, who, in August 1981, offered it as PC DOS 1.0 as the default operating system for the IBM PC.[]

Originally MS-DOS was designed to be an operating system that could run on any 8086-family computer. Each computer would have its own distinct hardware and its own version of MS-DOS, similar to the situation that existed for CP/M, and with MS-DOS emulating the same solution as CP/M to adapt for different hardware platforms. To this end, MS-DOS was designed with a modular structure with internal device drivers, minimally for primary disk drives and the console, integrated with the kernel and loaded by the boot loader, and installable device drivers for other devices loaded and integrated at boot time. The OEM would use a development kit provided by Microsoft to build a version of MS-DOS with their basic I/O drivers and a standard Microsoft kernel, which they would typically supply on disk to end users along with the hardware. Thus, there were many different versions of "MS-DOS" for different hardware, and there is a major distinction between an IBM-compatible (or ISA) machine and an MS-DOS [compatible] machine. Some machines, like the Tandy 2000, were MS-DOS compatible but not IBM-compatible, so they could only run software written exclusively for MS-DOS without dependence on the peripheral hardware of the IBM PC architecture.

This design would have worked well for compatibility, if application programs had only used MS-DOS services to perform device I/O, and indeed the same design philosophy is embodied in Windows NT (see Hardware Abstraction Layer). However, in MS-DOS's early days, the greater speed attainable by programs through direct control of hardware was of particular importance, especially for games, which often pushed the limits of their contemporary hardware. Very soon an IBM-compatible architecture became the goal, and before long all 8086-family computers closely emulated IBM's hardware, and only a single version of MS-DOS for a fixed hardware platform was needed for the market. This version is the version of MS-DOS that is discussed here, as the dozens of other OEM versions of "MS-DOS" were only relevant to the systems they were designed for, and in any case were very similar in function and capability to the same-numbered standard version for the IBM PC, with a few notable exceptions.

While MS-DOS appeared on PC clones, true IBM computers used PC DOS, a rebranded form of MS-DOS. Ironically, the dependence on IBM-compatible hardware caused major problems for the computer industry when the original design had to be changed. For example, the original design could support no more than 640 kilobytes of memory (the 640 KB barrier), because IBM's hardware design reserved the address space above this limit for peripheral devices and ROM. Manufacturers had to develop complicated schemes (EMS and XMS, and other minor proprietary ones) to access additional memory. This limitation would not have been a problem if the original idea of interfacing with hardware through MS-DOS had endured. (However, MS-DOS was also a real-mode operating system, and the Intel x86 architecture only supports up to 1 MB of memory address space in Real Mode, so for simple access to megabytes of memory, MS-DOS would have had to be rewritten to run in 80286 or 80386 Protected Mode.) Also, Microsoft originally described MS-DOS as "an operating system for Intel 8086-based microcomputers", and the 8086 CPU (and its cousin the 8088) itself has only 1 MB of total memory address space.

Versions

Microsoft licensed or released versions of MS-DOS under different names like Lifeboat Associates "Software Bus 86"[2] aka SB-DOS,[] COMPAQ-DOS,[2] NCR-DOS or Z-DOS[] before it eventually enforced the MS-DOS name for all versions but the IBM one, which was originally called "IBM Personal Computer DOS", later shortened to IBM PC DOS. (Competitors released compatible DOS systems such as DR DOS and PTS-DOS that could also run DOS applications.)

The following versions of MS-DOS were released to the public:[3] [4]

- MS-DOS 1.x
  - Version 1.12 (OEM) - Compaq release of PC DOS 1.10
  - Version 1.19 (OEM)[5] - Zenith OEM
  - Version 1.25 (OEM) - Microsoft repackaging of PC DOS 1.10
- MS-DOS 2.x - Support for 10 MB hard disk drives and tree-structure filing system
  - Version 2.0 (OEM)
  - Version 2.1 (OEM)
  - Version 2.11 (OEM)
- MS-DOS 3.x
  - Version 3.0 (OEM) - Support for FAT16
  - Version 3.1 (OEM) - Support for Microsoft Networks

- Version 3.2 (OEM)
- Version 3.21 (OEM)
- Version 3.25 (OEM)
- Version 3.3 (OEM)
- Version 3.3a (OEM)
- Version 3.31 (OEM) - Compaq MS-DOS 3.31 supports FAT16B and larger drives.[6]
- MS-DOS 4.x - includes a graphical/mouse interface. It had many bugs and compatibility issues.[7]
  - Version 4.00 (OEM)
  - Version 4.01 (OEM) - IBM patched Version 4.00 before Microsoft released it. First version to introduce volume serial number when formatting hard disks and floppy disks (Disk duplication also[8]).[9]
  - Version 4.01a (OEM)
- MS-DOS 5.x
  - Version 5.0 (Retail) - includes a full-screen editor. A number of bugs required re issue.
  - Version 5.0a (Retail) - With this release, IBM and Microsoft versions diverge.
  - Version 5.0.500 (WinNT) - All Windows NT 32-bit versions ship with files from DOS 5.0
- MS-DOS 6.x
  - Version 6.0 (Retail) - Online help through QBASIC. Disk compression and antivirus included.
  - Version 6.1 (none) - IBM and Microsoft alternate DOS 6 versions. IBM released 6.3 also.
  - Version 6.2 (Retail) - Scandisk as replacement for CHKDSK. Fix serious bugs in DBLSPACE.
  - Version 6.21 (Retail) - Stacker-infringing DBLSPACE removed.
  - Version 6.22 (Retail) - New DRVSPACE compression.
- MS-DOS 7.x
  - Version 7.0 (Win95, 95A) - Support for VFAT long file names and 32-bits signed integer errorlevel. New editor. JO.SYS is an alternative filename of the IO.SYS kernel file and used as such for "special purposes". JO.SYS allows booting from CD-ROM to hard disk.
  - Version 7.1 (Win95B - Win98SE) - Support for FAT32 file system. Last general purpose DOS to load Windows.
- MS-DOS 8.0
  - Version 8.0 (WinME) - Integrated drivers for faster Windows loading. Four different kernels (IO.SYS) observed.[10]
  - Version 8.0 (WinXP) - DOS boot disks created by XP and later contain files from WinME. The internal command prompt still reports version 5.0

Microsoft DOS was released through the OEM channel, until DRI released DR DOS 5.0 as a retail upgrade. With PC DOS 5.00.1, the IBM-Microsoft agreement started to end, and IBM entered the retail DOS market with IBMDOS 5.00.1, 5.02, 6.00 and PC DOS 6.10, 6.30, 7.00 and 2000.

A number of beta versions have surfaced on the Internet, such as 5.0 (a ten-diskette version in the same vein as 4.0), 7.00 beta 1 (based on 6.00), 7.00 beta 2 (based on 6.22). An OEM source package for 6.00, and a late release of 6.2(b) have also been seen. These are not retail versions.

Competition

The original MS-DOS advertisement in 1981.

On microcomputers based on the Intel 8086 and 8088 processors, including the IBM PC and clones, the initial competition to the PC DOS/MS-DOS line came from Digital Research, whose CP/M operating system had inspired MS-DOS. In fact, there remains controversy as to whether Q-DOS was more or less plagiarised from early versions of CP/M code. Digital Research released CP/M-86 a few months after MS-DOS, and it was offered as an alternative to MS-DOS and Microsoft's licensing requirements, but at a higher price. Executable programs for CP/M-86 and MS-DOS were not interchangeable with each other; much applications software was sold in both MS-DOS and CP/M-86 versions until MS-DOS became preponderant (later Digital Research operating systems could run both MS-DOS and CP/M-86 software). MS-DOS originally supported the simple .COM, which was modelled after a similar but binary incompatible format known from CP/M-80. CP/M-86 instead supported a relocatable format using the file extension .CMD to avoid name conflicts with CP/M-80 and MS-DOS .COM files. MS-DOS version 2.0 added a more advanced relocatable .EXE executable file format.

Most of the machines in the early days of MS-DOS had differing system architectures and there was a certain degree of incompatibility, and subsequently vendor lock-in. Users who began using MS-DOS with their machines were compelled to continue using the version customized for their hardware, or face trying to get all of their proprietary hardware and software to work with the new system.

In the business world the 808x-based machines that MS-DOS was tied to faced competition from the Unix operating system which ran on many different hardware architectures. Microsoft itself sold a version of Unix for the PC called Xenix.

In the emerging world of home users, a variety of other computers based on various other processors were in serious competition with the IBM PC: the Apple II, early Apple Macintosh, the Commodore 64 and others did not use the 808x processor; many 808x machines of different architectures used custom versions of MS-DOS. At first all these machines were in competition. In time the IBM PC hardware configuration became dominant in the 808x market as software written to communicate directly with the PC hardware without using standard operating system calls ran much faster, but on true PC-compatibles only. Non-PC-compatible 808x machines were too small a market to have fast software written for them alone, and the market remained open only for IBM PCs and machines that closely imitated their architecture, all running either a single version of MS-DOS compatible only with PCs, or the equivalent IBM PC DOS. Most clones cost much less than IBM-branded machines of similar performance, and became widely used by home users, while IBM PCs had a large share of the business computer market.

Microsoft and IBM together began what was intended as the follow-on to MS-DOS/PC DOS, called OS/2. When OS/2 was released in 1987, Microsoft began an advertising campaign announcing that "DOS is Dead" and stating that version 4 was the last full release. OS/2 was designed for efficient multi-tasking — an IBM speciality derived from deep experience with mainframe operating systems — and offered a number of advanced features that had been designed together with similar look and feel; it was seen as the legitimate heir to the "kludgy" DOS platform.

MS-DOS had grown in spurts, with many significant features being taken or duplicated from Microsoft's other products and operating systems. MS-DOS also grew by incorporating, by direct licensing or feature duplicating, the functionality of tools and utilities developed by independent companies, such as Norton Utilities, PC Tools (Microsoft Anti-Virus), QEMM expanded memory manager, Stacker disk compression, and others.

During the period when Digital Research was competing in the operating system market some computers, like Amstrad PC1512, were sold with floppy disks for two operating systems (only one of which could be used at a time), MS-DOS and CP/M-86 or a derivative of it. Digital Research produced DOS Plus, which was compatible with MS-DOS 2.11, supported CP/M-86 programs, had additional features including multi-tasking, and could read and write disks in CP/M and MS-DOS format.

While OS/2 was under protracted development, Digital Research released the MS-DOS compatible DR DOS 5.0, which included features only available as third-party add-ons for MS-DOS. Unwilling to lose any portion of the market, Microsoft responded by announcing the "pending" release of MS-DOS 5.0 in May 1990. This effectively killed most DR DOS sales until the actual release of MS-DOS 5.0 in June 1991. Digital Research brought out DR DOS 6.0, which sold well until the "pre-announcement" of MS-DOS 6.0 again stifled the sales of DR DOS.

Microsoft had been accused of carefully orchestrating leaks about future versions of MS-DOS in an attempt to create what in the industry is called FUD (fear, uncertainty, and doubt) regarding DR DOS. For example, in October 1990, shortly after the release of DR DOS 5.0, and long before the eventual June 1991 release of MS-DOS 5.0, stories on feature enhancements in MS-DOS started to appear in *InfoWorld* and *PC Week*. Brad Silverberg, Vice President of Systems Software at Microsoft and General Manager of its Windows and MS-DOS Business Unit, wrote a forceful letter to *PC Week* (November 5, 1990), denying that Microsoft was engaged in FUD tactics ("to serve our customers better, we decided to be more forthcoming about version 5.0") and denying that Microsoft copied features from DR DOS:

> "The feature enhancements of MS-DOS version 5.0 were decided and development was begun long before we heard about DR DOS 5.0. There will be some similar features. With 50 million MS-DOS users, it shouldn't be surprising that DRI has heard some of the same requests from customers that we have." – (Schulman et al. 1994).[]

The pact between Microsoft and IBM to promote OS/2 began to fall apart in 1990 when Windows 3.0 became a marketplace success. Much of Microsoft's further contributions to OS/2 also went into creating a third GUI replacement for DOS, Windows NT.

IBM, which had already been developing the next version of OS/2, carried on development of the platform without Microsoft and sold it as the alternative to DOS and Windows.

Legal issues

As a response to Digital Research's DR DOS 6.0, which bundled SuperStor disk compression, Microsoft opened negotiations with Stac Electronics, vendor of the most popular DOS disk compression tool, Stacker. In the due diligence process, Stac engineers had shown Microsoft part of the Stacker source code. Stac was unwilling to meet Microsoft's terms for licensing Stacker and withdrew from the negotiations. Microsoft chose to license Vertisoft's DoubleDisk, using it as the core for its DoubleSpace disk compression.[11]

MS-DOS 6.0 and 6.20 were released in 1993, both including the Microsoft DoubleSpace disk compression utility program. Stac successfully sued Microsoft for patent infringement regarding the compression algorithm used in DoubleSpace. This resulted in the 1994 release of MS-DOS 6.21, which had disk-compression removed. Shortly afterwards came version 6.22, with a new version of the disk compression system, DriveSpace, which had a different compression algorithm to avoid the infringing code.

Prior to 1995, Microsoft licensed MS-DOS (and Windows) to computer manufacturers under three types of agreement: per-processor (a fee for each system the company sold), per-system (a fee for each system of a particular model), or per-copy (a fee for each copy of MS-DOS installed). The largest manufacturers used the per-processor arrangement, which had the lowest fee. This arrangement made it expensive for the large manufacturers to migrate to any other operating system, such as DR DOS. In 1991, the U.S. government Federal Trade Commission began investigating Microsoft's licensing procedures, resulting in a 1994 settlement agreement limiting Microsoft to per-copy licensing. Digital Research did not gain by this settlement, and years later its successor in interest, Caldera, sued Microsoft for damages. It was believed that the settlement ran in the order of $150m, but was revealed in November 2009 with the release of the Settlement Agreement to be $280m.

Use of undocumented APIs

Microsoft also used a variety of tactics in MS-DOS and several of their applications and development tools that, while operating perfectly when running on genuine MS-DOS (and PC DOS), would break when run on another vendor's implementation of DOS. Notable examples of this practice included:

- Microsoft's QuickPascal released in early 1989 was the first MS product that checked for MS-DOS by modifying the program's Program Segment Prefix using undocumented DOS functions, and then checked whether or not the associated value changed in a fixed position within the DOS data segment (also undocumented). This check also made it into later MS products, including Microsoft QuickC v2.5, Programmer's Workbench and Microsoft C v6.0.[]
- The (once infamous) AARD code, a block of code in the Windows 3.1 beta installer. It was XOR encrypted, self-modifying, and deliberately obfuscated, using various undocumented DOS structures and functions to determine whether or not Windows really was running on MS-DOS.[]
  - Note that the Windows 3.0 beta code only gave a warning that Windows would not operate properly on a "foreign" OS. It did in fact run just fine on DR DOS 6.0.
- Interrupt routines called by Windows to inform MS-DOS that Windows is starting/exiting, information that MS-DOS retained in an IN_WINDOWS flag, in spite of the fact that MS-DOS and Windows were supposed to be two separate products.[]

End of MS-DOS

In 2011, MS-DOS is still used in some enterprises to run legacy applications, such as this US Navy food service management system.

Since the release of Windows 95 up until Windows NT, it was segregrated as a full product used for bootstrapping, troubleshooting, and backwards-compatibility with old DOS games and no longer released as a standalone product.

Windows XP contains a copy of the Windows Me boot disk, stripped down to bootstrap only. This is accessible only by formatting a floppy as an "MS-DOS startup disk". Files like the driver for the CD-ROM support were deleted from the Windows ME bootdisk and the startup files (AUTOEXEC.BAT and CONFIG.SYS) no longer had a content. This modified disk was the base for creating the MS-DOS image for Windows XP. Some of the deleted files can be recovered with an undelete tool.[12] With Windows Vista the files on the startup disk are dated 18 April 2005 but are otherwise unchanged, including the string "MS-DOS Version 8 © Copyright 1981-1999 Microsoft Corp" inside COMMAND.COM.

However the only versions of DOS currently recognized as stand-alone OSs, and supported as such by the Microsoft Corporation are DOS 6.0 and 6.22, both of which remain available for download via their MSDN, volume license, and OEM license partner websites, for customers with valid login credentials.

MS-DOS is still used in embedded x86 systems due to its simple architecture and minimal memory and processor requirements. The command line interpreter of NT-based versions of Windows, cmd.exe, maintains most of the same commands and some compatibility with DOS batch files.

Windows command-line interface

All versions of Microsoft Windows have had an MS-DOS like command-line interface (CLI). This could run many DOS and variously Win32, OS/2 1.x and Posix command line utilities in the same command-line session, allowing piping between commands. The user interface, and the icon up to Windows 2000, followed the native MS-DOS interface.

The 16-bit versions of Windows (up to 3.11) ran as a Graphical User Interface (GUI) on top of MS-DOS. With Windows 95, 98, and ME the MS-DOS part was integrated, treating both operating systems as a complete package. The command line accessed the DOS command line (usually command.com), through a Windows module (winoldap.mod).

A new line of Windows, (Windows NT), boot through a kernel whose sole purpose is to load Windows. One can not run Win32 applications in the loader system in the manner that OS/2, UNIX or Consumer Windows can launch character mode sessions.

The command session permits running of various supported command line utilities from Win32, MS-DOS, OS/2 1.x and POSIX. The emulators for MS-DOS, OS/2 and POSIX use the host's window in the same way that Win16 applications use the Win32 explorer. Using the host's window allows one to pipe output between emulations.

The MS-DOS emulation is done through the NTVDM (NT Virtual DOS Machine). This is a modified SoftPC (a former product similar to VirtualPC), running a modified MS-DOS 5 (ntio.sys and ntdos.sys). The output is handled by the console DLLs, so that the program at the prompt (cmd.exe, 4nt.exe, tcc.exe), can see the output. Command.com passes most of its commands down to the underlying processor (cmd.exe, 4nt.exe, ...), but is generally not used. If you need commands that rely on TSR programs, you must launch these from command.com. 64-bit Windows does not have either the DOS emulation, or the DOS commands (edit, debug, edlin, kbd16), that come with 32-bit Windows.

The DOS version returns 5.00 or 5.50, depending on which interrupt is used to determine it. Utilities from MS-DOS 5.00 run in this emulation without modification. The very early beta programs of NT show MS-DOS 30.00, but programs running in MS-DOS 30.00 would assume that OS/2 was in control.

The OS/2 emulation is handled through OS2SS.EXE and OS2.EXE, and DOSCALLS.DLL. OS2.EXE is a version of the OS/2 shell (cmd.exe), which passes commands down to the OS2SS.EXE, and input-output to the Windows NT shell. Windows 2000 was the last version of NT to support OS/2. The emulation is OS/2 1.30.

POSIX is emulated through the POSIX shell, but no emulated shell: the commands are handled directly in CMD.EXE.

The Command Prompt is often called the MS-DOS prompt. In part, this was the official name for it in Windows 9x and early versions of Windows NT (NT 3.5 and earlier), and in part because the SoftPC emulation of DOS redirects output into it. Some purists object to this name, because the underlying binaries are Win32 console applications, rather than DOS programs. This confusion does not exist under OS/2 because there are separate DOS and OS/2 prompts, and running a DOS program under OS/2 will launch a separate DOS window to run the application.

All versions of Windows for x86-64 and Itanium architectures no longer include the NTVDM and can therefore no longer natively run MS-DOS or 16-bit Windows applications. There are alternatives in the form of Virtual machine emulators such as Microsoft's own Virtual PC, as well as VMware, DOSBox, and others.

Legacy compatibility

From 1983 onwards, various companies worked on graphical user interfaces (GUIs) capable of running on PC hardware. With DOS being the dominant operating system several companies released alternate shells, e.g. Microsoft Word for DOS, XTree, and the Norton Commander. However, this required duplication of effort and did not provide much consistency in interface design (even between products from the same company).

Later, in 1985, Microsoft Windows was released as Microsoft's first attempt at providing a consistent user interface (for applications). The early versions of Windows ran on top of MS-DOS and its clones. At first Windows met with little success, but this was also true for most other companies' efforts as well, for example GEM. After version 3.0 (1990), Windows gained market acceptance.

Windows 9x used the DOS boot process to launch into protected mode. Basic features related to the file system, such as long file names, were only available to DOS when running as a subsystem of Windows. Windows NT runs independently of DOS but includes NTVDM, a component for simulating a DOS environment for legacy applications.

Related systems

Single-user

Several similar products were produced by other companies. In the case of PC DOS and DR DOS, it is common but incorrect to call these "clones". Given that Microsoft manufactured PC DOS for IBM, PC DOS and MS-DOS were (to continue the genetic analogy) "identical twins" that diverged only in adulthood and eventually became quite different products. Although DR DOS is regarded as a clone of MS-DOS, the DR DOS versions appeared months and years before Microsoft's products. (For example, MS-DOS 4, released in July 1988, was followed by DR DOS 5.0 in May 1990. MS-DOS 5.0 came in April 1991, with DR DOS *6.0* being released the following June. MS-DOS 6.0 did not arrive until April 1993, with Novell DOS 7, DR DOS' successor, following the next month.[13]) What made the difference in the end was Microsoft's control of the Windows platform and their programming practices which intentionally made Windows appear as if it ran poorly on competing versions of DOS.[] DRI (DOS 6 update) had to release interim releases to circumvent Windows limitations inserted artificially,[] designed specifically to provide Microsoft with an unfair competitive advantage.[]

- PC DOS,
- DR DOS / Novell DOS / OpenDOS,
- GNU/DOS / FreeDOS / FreeDOS 32,
- PTS-DOS,
- ROM-DOS.

These products are collectively referred to as DOS. However, MS-DOS can be a generic reference to DOS on IBM-PC compatible computers.

See also

- 4DOS - designed to replace the default command interpreter COMMAND.COM
- Bad command or file name
- Comparison of DOS operating systems
- DOSBox
- DOSKey, MS-DOS utility
- History of Microsoft Windows
- List of DOS commands
- List of Microsoft Windows versions
- List of MS-DOS games
- Microsoft Windows
- MS-DOS API
- Tao ExDOS - Commercial solution for DOS software on new Windows systems.
- Timeline of DOS operating systems
- Win32 console - a text-rendering system akin to MS-DOS

Notes

[2] Ray Duncan (1988). *The MS-DOS Encyclopedia - version 1.0 through 3.2.* Microsoft Press. ISBN 1-55615-049-0.
[6] Up to 512 MiB only.
[7] http://www.techrepublic.com/photos/dinosaur-sightings-installing-ms-dos-4/36455?mobileView=fullonce
[8] Only if boot record of source floppy disk contains volume serial number also.
[10] One for the floppy disk, one for a bootable partition of a hard disk, the other two are not known.
[11] BYTE Magazine, How Safe is Disk Compression? (http://www.byte.com/art/9402/sec6/art1.htm), February, 1994.
[12] (http://web.archive.org/web/20101024145636/http://mirror.href.com/thestarman/winxp/winxpsd.htm)
[13] Timeline 1980-1993 (http://www.oldfiles.org.uk/powerload/timeline.htm)

References

- Microsoft MS-DOS 6 Technical Reference (http://technet.microsoft.com/en-us/library/cc743186.aspx)

External links

- MS-DOS Resources (http://www.josephn.net/ms-dos) - Information on various aspects of MS-DOS including downloads
- Current License Agreement Policies for MS-DOS and Windows (http://support.microsoft.com/kb/79747)
- Tim Paterson on DOS (http://www.patersontech.com/Dos/Articles.aspx) - Paterson wrote the QDOS OS
- MS-DOS: A Brief Introduction (http://www.linfo.org/ms-dos.html)
- Richard Bonner's DOS website (http://www.chebucto.ns.ca/~ak621/DOS/DOS-Head.html)
- Batfiles (http://purl.oclc.org/net/Batfiles/) - the DOS batch file programming handbook and tutorial
- Arachne graphical browser for DOS (http://browser.arachne.cz/)
- Linux/dosemu (http://www.dosemu.org/)
- *Ralf Brown's Interrupt List* (http://www-2.cs.cmu.edu/afs/cs/user/ralf/pub/WWW/files.html)
- DOS command overview (http://www.computerhope.com/msdos.htm)
- DOSBox, a multiplatform DOS emulator (http://dosbox.sourceforge.net/news.php?show_news=1)
- Garbo - An MS-DOS program distribution library (http://garbo.uwasa.fi/) at the University of Vaasa, Finland
- MS-DOS 6 Technical Reference at Microsoft TechNet (http://www.microsoft.com/technet/archive/msdos/01_intro.mspx?mfr=true)
- Promotional video for MS-DOS 5 (http://www.youtube.com/watch?v=dmEvPZUdAVI)

- List of all released Microsoft Operating Systems and betas (http://69.117.216.225:81/OSArchive/microsoft.htm)

Windows_95

Windows 95 is a consumer-oriented graphical user interface-based operating system. It was released on August 24, 1995 by Microsoft,[] [1] and was a significant progression from the company's previous Windows products. During development, it was referred to as **Windows 4.0** or by the internal codename **Chicago**.

Windows 95 integrated Microsoft's formerly separate MS-DOS and Windows products. It featured significant improvements over its predecessor, Windows 3.1, most notably in the graphical user interface (GUI) and in its relatively simplified "plug-n-play" features. There were also major changes made at lower levels of the operating system, such as moving from a mainly 16-bit architecture to a pre-emptively multitasked 32-bit architecture.

Accompanied by an extensive marketing campaign, Windows 95 was a major success in the marketplace at launch and shortly became the most popular desktop operating system. It also introduced numerous functions and features that were featured in later Windows versions, such as the taskbar, the 'Start' button, and the way the user navigates. It was also suggested that Windows 95 had an effect of driving other major players (including OS/2) out of business, something which would later be used in court against Microsoft.

Some three years after its introduction, Windows 95 was succeeded by Windows 98. Support for Windows 95 ended on December 31, 2001.

Development

The initial design and planning of Windows 95 can be traced back to around March 1992,[2] [3] [4] just after the release of Windows 3.1. At this time, *Windows for Workgroups 3.1* and Windows NT 3.1 were still in development and Microsoft's plan for the future was focused on Cairo. Cairo would be Microsoft's next-generation operating system based on Windows NT and featuring a new user interface and an object-based file system, but it was not planned to be shipped before 1994 (Cairo would eventually partially ship in July 1996 in the form of Windows NT 4.0, but without the object-based file system, which would later evolve into WinFS).

Simultaneously with Windows 3.1's release, IBM started shipping OS/2 2.0. Microsoft realized they were in need of an updated version of Windows that could support 32-bit applications and preemptive multitasking, but could still run on low-end hardware (Windows NT did not). So the development of Windows "Chicago" was started and, as it was planned for a late 1993 release, became known as Windows 93. Initially, the decision was made not to include a new user interface, as this was planned for Cairo, and only focus on making installation, configuration, and networking easier. Windows 93 would ship together with MS-DOS 7.0, offering a more integrated experience to the user and making it pointless for other companies to create DOS clones. MS-DOS 7.0 was in development at that time under the code name "Jaguar" and could optionally run on top of a Windows 3.1-based 32-bit protected mode kernel called "Cougar" in order to better compete with DR-DOS. The first version of Chicago's feature specification was finished on September 30, 1992. Cougar was to become Chicago's kernel.

Beta

Prior to the official release, the American public was given a chance to preview Windows 95 in the Windows 95 Preview Program. For US$19.95, users were sent a set of 3.5-inch floppy diskettes that would install Windows 95 either as an upgrade to Windows 3.1x or as a fresh install on a clean computer. Users who bought into the program were also given a free preview of The Microsoft Network (MSN), the online service that Microsoft launched with Windows 95. During the preview period Microsoft established various electronic distribution points for promotional and technical documentation on Chicago[5] including a detailed document for media reviewers describing the new system highlights.[5] [6] The preview versions expired in November 1995, after which the user would have to purchase their own copy of the final version of Windows 95. Some releases were made that haven't been leaked, such as Chicago 30.

Release

Windows 95 was released with great fanfare, including a commercial featuring the Rolling Stones' 1981 single "Start Me Up" (a reference to the Start button).[7] It was widely reported that Microsoft paid the Rolling Stones between US$8 and US$14 million for the use of the song in the 95 advertising campaign. According to sources at Microsoft, however, this was just a rumor spread by the band to increase their market value, and Microsoft actually paid a fraction of that amount.[8] A 30-minute promotional video, labeled a "cyber sitcom", featuring Jennifer Aniston and Matthew Perry, was also released to showcase the features of Windows 95. Microsoft's US$300 million advertising campaign featured stories of people waiting in line outside stores to get a copy.

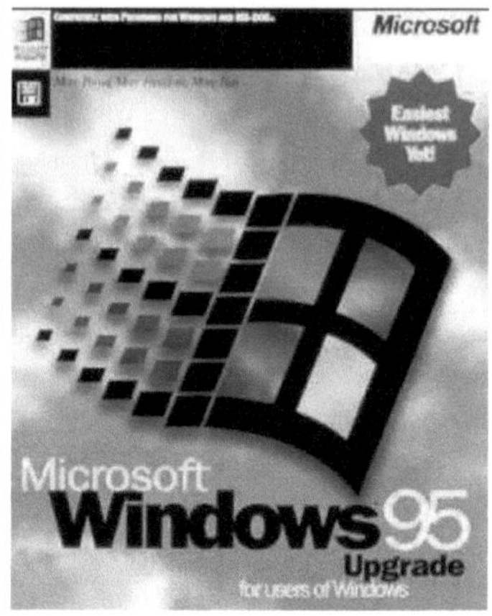

Microsoft Windows 95 operating system cover shot

In the UK, the largest computer chain PC World received a large number of oversized Windows 95 boxes, posters and point of sale material, and many branches opened at midnight to sell the first copies of the product, although these customers were far fewer in number than publicity had suggested. In London, Microsoft gave free newspapers to people.[]

In the United States, the Empire State Building in New York City was lit to match the colors of the Windows logo.[] In Canada, a 328 ft (m) banner was hung from the top of the CN Tower in Toronto. Copies of The Times were available for free in the United Kingdom where Microsoft paid for 1.5 million issues (twice the daily circulation at the time).

The release included a number of "Fun Stuff" items on the CD, including music videos of Edie Brickell's "Good Times"[9] and Weezer's "Buddy Holly". On the CD version a computer game called Hover! was also included.

Architecture

Windows 95 was designed to be maximally compatible with existing MS-DOS and 16-bit Windows programs and device drivers, while offering a more stable and better performing system.[] [] Windows 95 architecture is an evolution of Windows for Workgroups' 386 enhanced mode. The lowest level of the operating system consists of a large number of *virtual device drivers* (VxDs) running in 32-bit protected mode and one or more virtual DOS machines running in virtual 8086 mode. The virtual device drivers are responsible for handling physical devices (such as video and network cards), emulating virtual devices used by the virtual machines, or providing various system services. The three most important virtual device drivers are:

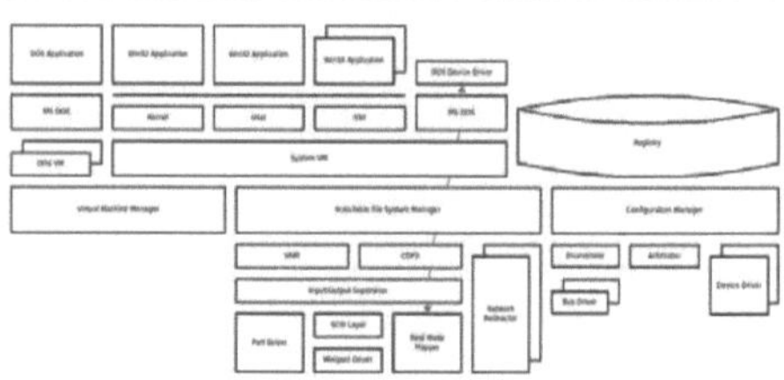
Architectural diagram

Virtual Machine Manager (VMM32.VXD)

Responsible for memory management, event handling, interrupt handling, loading and initializing virtual device drivers, creating new virtual machines and thread scheduling

Configuration Manager (CONFIGMG)

Responsible for implementing Plug and Play functionality, monitoring hardware configuration changes, detecting devices using *bus enumerators*, and allocating I/O ports, IRQs, DMA channels and memory in a conflict free fashion

Installable File System Manager (Input/Output Subsystem)

Coordinates access to supported file systems. Windows 95 initially shipped with support for FAT12, FAT16, VFAT extension, ISO 9660 (CDFS) and network redirectors. Later releases included support for FAT32.

Access requests to physical media are sent to *Input/Output Supervisor*, a component responsible for scheduling the requests. Each physical media has its own device driver: Access to the disk is performed by a *port driver*, while access to a SCSI device is handled by a *miniport* driver working atop the SCSI layer. Port and miniport drivers perform I/O operations in 32-bit protected mode, bypassing MS-DOS and BIOS, giving a significant performance improvement. In case there is no native Windows driver for a certain storage device, or if a device is forced to run in compatibility mode, the *Real Mode Mapper* can access it through MS-DOS.

32-bit Windows programs are assigned their own memory segments, which can be adjusted to any size the user wishes. Memory area outside the segment cannot be accessed by a program. If they crash, they do not harm anything else. Before this, programs used fixed non-exclusive 64 KB segments. While the 64 KB size was a serious handicap in DOS and Windows 3.x, lack of guarantee of exclusiveness was the cause of stability issues because programs sometimes overwrote each other's segments. A crashing Windows 3.x program could knock out surrounding processes.

The Win32 API is implemented by three modules, each consisting of a 16-bit and a 32-bit component:

Kernel

Provides high level access to memory and process management, and access to the file system. Consists of KRNL386.EXE, KERNEL32.DLL, and VWIN32.VXD.

User

Responsible for managing and drawing the various user interface components, such as windows, menus and buttons. Consists of USER.EXE and USER32.DLL.

Graphics Device Interface (GDI)

Responsible for drawing graphics in a device-independent way. Consists of GDI.EXE and GDI32.DLL.

Dependence on MS-DOS

To end-users, MS-DOS appears as an underlying component of Windows 95. For example, it is possible to prevent loading the graphical user interface and boot the system into a real-mode MS-DOS environment. This sparked debate amongst users and professionals over the question of to what extent Windows 95 is an operating system or merely a graphical shell running on top of MS-DOS.[] [] []

When the graphical user interface is started, the virtual machine manager takes over the filesystem-related and disk-related functionality. MS-DOS itself is demoted to a compatibility layer for 16-bit device drivers.[] This contrasts with earlier versions of Windows which relies on MS-DOS to perform file and disk access (Windows for Workgroups 3.11 could also largely bypass MS-DOS when 32-bit file access and 32-bit disk access were enabled). Keeping MS-DOS in memory allows Windows 95 to use DOS device drivers when suitable Windows drivers are unavailable. Windows 95 is capable of using all 16-bit Windows 3.x drivers.

Contrary to Windows 3.1x, DOS programs running in Windows 95 do not need DOS drivers for mouse, CD-ROM access and sound card; Windows drivers are used instead. HIMEM.SYS is still required to boot Windows 95. EMM386 and other memory managers, however, are only used by legacy DOS programs. In addition, CONFIG.SYS and AUTOEXEC.BAT settings (aside from HIMEM.SYS) have no effect on Windows programs. DOS games, which could not be executed on Windows 3.x, can run inside Windows 95. (Games tended to lock up Windows 3.x or cause other problems). As with Windows 3.x, DOS programs that use EGA or VGA graphics modes run in windowed mode (CGA and text mode programs can continue to run).

On startup, the MS-DOS component in Windows 95 responds to a pressed <kbd class="keyboard-key nowrap" style="border: 1px solid #aaa; -moz-border-radius: 2px; -webkit-border-radius: 2px; border-radius: 2px; -moz-box-shadow: 1px 2px 2px #ddd; -webkit-box-shadow: 1px 2px 2px #ddd; box-shadow: 1px 2px 2px #ddd; background-color: #f9f9f9; background-image: -moz-linear-gradient(top, #eee, #f9f9f9, #eee); background-image: -ms-linear-gradient(top, #eee, #f9f9f9, #eee); background-image: -o-linear-gradient(top, #eee, #f9f9f9, #eee); background-image: -webkit-linear-gradient(top, #eee, #f9f9f9, #eee); background-image: linear-gradient(top, #eee, #f9f9f9, #eee); padding: 1px 3px; font-family: inherit; font-size: 0.85em;">F8</kbd> key by temporarily pausing the default boot process and presenting the DOS boot options menu, allowing the user to continue starting Windows normally, start Windows in safe mode or exit to the DOS prompt.[] As in previous versions of MS-DOS, there is no 32-bit support and DOS drivers must be loaded for mice and other hardware.

As a consequence of being DOS-based, Windows 95 has to keep internal DOS data structures synchronized with those of Windows 95. When starting a program, even a native 32-bit Windows program, MS-DOS momentarily executes to create a data structure known as the Program Segment Prefix. It is even possible for MS-DOS to run out of conventional memory while doing so, preventing the program from launching.[] Windows 3.x allocated *fixed* segments in conventional memory first. Since the segments were allocated as fixed, Windows could not move them, which would prevent any more programs from launching. This was circumvented in Windows 95.

Microsoft partially removed support for the File Control Blocks (an API hold-over of DOS 1.x and CP/M) in Windows 95 OSR 2. FCB functions cannot write to FAT32 volumes, only read them.

User interface

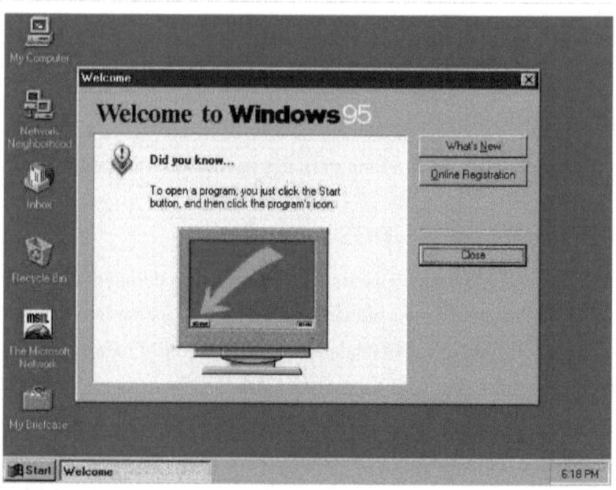

Windows 95's new interface was introduced by a "Welcome" dialog on its first startup.

Windows 95 introduced a redesigned shell based around a desktop metaphor; unlike Windows 3.1 (where it was used to display running applications), the desktop was re-purposed to hold shortcuts to applications, files, and folders. Running applications were now displayed as buttons on a taskbar across the bottom of the screen, which also contains a notification area used to display icons for background applications, a volume control, and the current time. The Start menu, invoked by clicking the "Start" button also contained on the taskbar, was introduced as an additional means of launching applications or opening documents—while maintaining the program groups used by its predecessor, Program

Manager, it now displayed applications within cascading sub-menus. The previous File Manager program was also replaced by Windows Explorer.

In 1994, Microsoft corporation designers Mark Malamud and Erik Gavriluk approached Brian Eno to compose music for the Windows 95 project.[10] The result was the six-second start-up music-sound of the Windows 95 operating system, *The Microsoft Sound*.[11]

When released for Windows 95 and NT4, Internet Explorer 4 came with an optional Windows Desktop Update, which modified the shell to provide new features integrated with Internet Explorer, such as Active Desktop (which allowed internet content to be displayed directly on the desktop) and additional updates to Windows Explorer.

Some of the user interface elements introduced in Windows 95—such as the desktop, taskbar, Start menu, and Windows Explorer file manager, remained fundamentally unchanged on future versions of Windows.

Technical improvements

Windows 95 included support for 255-character mixed-case long filenames[12] and preemptively multitasked protected-mode 32-bit applications.

Long file names

32-bit File Access is necessary for the *long file names* feature introduced with Windows 95 through the use of the VFAT file system extension. It is available to both Windows programs and MS-DOS programs started from Windows (they have to be adapted slightly, since accessing long file names requires using larger pathname buffers and hence different system calls). Competing DOS-compatible operating systems released before Windows 95 cannot see these names. Using older versions of DOS utilities to manipulate files means that the long names are not visible and are lost if files are moved or renamed, as well as by the copy (but not the original), if the file is copied. During a Windows 95 automatic upgrade of an older Windows 3.1 system, DOS and third-party disk utilities which can destroy long file names are identified and made unavailable (Microsoft Anti-Virus for Windows indicated that the upgrade program was itself a computer virus). When Windows 95 is started in DOS mode, e.g. for running DOS programs, low-level access to disks is locked out. In case the need arises to depend on disk utilities that do not recognize long file names, such as MS-DOS 6.x's defrag utility, a program called LFNBACK for backup and restoration of long file names is provided on the CD-ROM. The program is in the \ADMIN\APPTOOLS\LFNBACK directory of the Windows 95 CD-ROM.

32-bit

Windows 95 followed Windows for Workgroups 3.11 with its lack of support for older, 16-bit x86 processors, thus requiring an Intel 80386 (or compatible). While the OS kernel is 32-bit, much code (especially for the user interface) remained 16-bit for performance reasons as well as development time constraints (much of Windows 95's UI code was recycled from Windows 3.1). This had a rather detrimental effect on system stability and led to frequent application crashes.

The introduction of 32-bit File Access in Windows for Workgroups 3.11 meant that 16-bit real mode MS-DOS is not used for managing the files while Windows is running, and the earlier introduction of the 32-bit Disk Access means that the PC BIOS is often no longer used for managing hard disks. DOS can be used for running old-style drivers for compatibility, but Microsoft discourages using them, as this prevents proper multitasking and impairs system stability. Control Panel allows a user to see what MS-DOS components are used by the system; optimal performance is achieved when they are bypassed. The Windows kernel uses MS-DOS style real-mode drivers in *Safe Mode,* which exists to allow a user to fix problems relating to loading native, protected-mode drivers.

Internet Explorer

Windows 95 originally shipped without Internet Explorer, and the default network installation did not install TCP/IP, the network protocol used on the Internet. At the release date of Windows 95, Internet Explorer 1.0 was available, but only in the Plus! add-on pack for Windows 95, which was a separate product. The Plus! Pack did not reach as many retail consumers as the operating system itself (it was mainly advertised for its add-ons such as themes and better disk compression) but was usually included in pre-installed (OEM) sales, and at the time of Windows 95 release, the web was being browsed mainly with a variety of early web browsers such as Netscape (promoted by products such as Internet in a Box).

Windows 95 OEM Service Release 1 was the first release of Windows to include Internet Explorer (version 2.0) with the OS. While there was no uninstaller, it could be deleted easily if the user so desired. OEM Service Release 2 included Internet Explorer 3. The installation of Internet Explorer 4 on Windows 95 (or the OSR2.5 version preinstalled on a computer) gave Windows 95 active desktop and browser integration into Windows Explorer, known as the Windows Desktop Update. The CD version of the last release of Windows 95, OEM Service Release 2.5 (Version 4.00.950C), includes Internet Explorer 4, and installs it after Windows 95's initial setup and first boot is complete.

Only the 4.x series of the browser contained the Windows Desktop Update features, so anyone wanting the new shell had to install IE4 with the desktop update before installing a newer version of Internet Explorer. The last version of Internet Explorer supported on Windows 95 is Internet Explorer 5.5 which was released in 2000. Windows 95 shipped with Microsoft's own dial-up online service called The Microsoft Network.

Editions

A number of different editions of Windows 95 have been released. Only the original release was sold as a shrink-wrapped product, later editions were provided only to computer original equipment manufacturers (OEMs) for installation on new PCs. For this reason these editions are known as **OEM Service Releases** (**OSR**).

Together with the introduction of Windows 95, Microsoft released the *Microsoft Plus! for Windows 95* pack, which contained a number of optional components for high-end (486) multimedia PCs, including Internet Explorer, DriveSpace, and additional themes.

Microsoft initially indicated to make updates available to Windows 95 every 6 months in the form of service packs. The growing availability of Internet access meant that Windows updates could now be downloaded from Microsoft directly. The first service pack was made available half a year after the original release and fixed a number of small bugs.

The second service pack mainly introduced support for new hardware. Most notably support for hard drives larger than 2 GB in the form of the FAT32 file system. This release was never made available to end-users directly and was only sold through OEMs with the purchase of a new PC.

A full third service pack was never released, but two smaller updates to the second were released in the form of a USB Supplement (OSR 2.1) and the Windows Desktop Update (OSR 2.5). Both were available as stand-alone updates and as updated disc images shipped by OEMs. OSR 2.5 was notable for featuring a number of changes to the Windows Explorer, integrating it with Internet Explorer 4.0—this version of the Explorer looks very similar to the one featured in Windows 98.

ease	Code name	Release date	Version			Software components				Hardware support							
			System properties[13]	System files[14]	Timestamp	MS-DOS	Internet Explorer	DriveSpace	DirectX[15]	FAT32	Infrared[16]	UDMA[17]	IRQ steering[18]	USB	AGP	MMX[19]	P6[20] [21]
dows retail OEM)	Chicago	24 August 1995	4.00.950		1995-07-11 09:50:00	7.0	N/A	2	N/A	No	No	No	No	No	No	Bugs	Bugs
rosoft s! for dows 95	Frosting		N/A	4.40.310	1995-07-14 04:40:00		1.0	3									
rvice ck 1	N/A	14 February 1996[22]	4.00.950a	4.00.951[23]	1995-12-31 09:50:00		2.0[24]	2			Yes						
EM rvice ase 1					1996-02-02 09:51:00												
EM rvice ase 2	Detroit	24 August 1996	4.00.950 B	4.00.1111	1996-08-24 11:11:11	7.1	3.0	3	2.0a	Yes		Yes	Yes			Yes	
SB lement OSR2	N/A	27 August 1997[25]		4.03.1212[26] 4.03.1214[27]	1997-04-10 12:14:00									Yes	Yes		
EM rvice ase 2.1																	
EM rvice ase 2.5		26 November 1997	4.00.950 C	4.03.1216[28]	1997-11-26 12:16:00		4.00		5.0								Yes

[2] Comes v. Microsoft. Plaintiff's Exhibit 1263 (http://groklaw.net/pdf/iowa/www.iowaconsumercase.org/011607/1000/PX01263.pdf).

[3] Comes v. Microsoft. Plaintiff's Exhibit 1308 (http://groklaw.net/pdf/iowa/www.iowaconsumercase.org/011607/1000/PX01308.pdf).

[4] Comes v. Microsoft. Plaintiff's Exhibit 1310 (http://groklaw.net/pdf/iowa/www.iowaconsumercase.org/011607/1000/PX01310.pdf).

[5] Microsoft Windows Chicago Reviewer's Guide (http://www.anonymous-insider.net/windows/research/acrobat/940601.pdf), p.282

[7] Microsoft detractors were quick to point out that the second verse of "Start Me Up" begins "you make a grown man cry" (a line which is repeated throughout). The phrase subsequently featured as a humorous reference in many critical expositions of Windows 95.

[8] Internet Archive

[12] 255-character mixed-case long filename is possible only for files and/or folders with no sub-folders at the root folder of any drive.

[13] The version string displayed in the "System properties" tab. Right-click on "My Computer" and choose "Properties".

[14] The version of updated system files. Note that most system files which have not been updated often retain their old version number. Version numbers are not consistently used: some system files may have older or newer build numbers or use a version numbering scheme separate from regular system files.

[15] Upgradable to 8.0a.

[23] Some components have higher build numbers up to 955.

[26] Original release of the USB Supplement to OSR2.

[27] Updated version of the USB Supplement to OSR2.

[28] The Microsoft Knowledge Base reports 4.03.1214. The USB Supplement to OSR2 contains an updated VMM.VXD with support for the Pentium Pro and Pentium II. This file has version 4.03.1216 and has a timestamp of 1997-09-23 09:51:18.

System requirements

Official system requirements were an Intel 80386 DX CPU of any speed, 4 MB of system RAM, and 50–55 MB of hard drive space depending on features selected. These minimal claims were made in order to maximize the available market of Windows 3.1 converts. This configuration was distinctly suboptimal for any productive use on anything but single tasking dedicated workstations due to the heavy reliance on virtual memory. Also, in some cases, if any networking or similar components were installed the system would refuse to boot with 4 megabytes of RAM. It was possible to run Windows 95 on a 386 SX but this led to even less acceptable performance due to its 16-bit external data bus. To achieve optimal performance, Microsoft recommends an Intel 80486 or compatible microprocessor with at least 8 MB of RAM.[1] Windows 95 fails to boot on computers with approximately more than 480 MB of memory.[2]

Windows 95 was superseded by Windows 98 and could still be directly upgraded by both Windows 2000 Professional edition[3] and Windows ME. On December 31, 2001, Microsoft ended its support for Windows 95, making it an "obsolete" product according to the Microsoft Lifecycle Policy.[4] Even though support for Windows 95 has ended, the software has occasionally remained in use on legacy systems for various purposes. In addition, some video game enthusiasts choose to use Windows 95 for their legacy system to play old DOS games, although some other versions of Windows such as Windows 98 can also be used for this purpose.

Most copies of Windows 95 were on CD-ROM, but a floppy version could also be had for older machines. The retail floppy disk version of Windows 95 came on 13 DMF formatted floppy disks, while OSR 2.1 doubled the floppy count to 26. Both versions exclude additional software that CD-ROM might have featured. Microsoft Plus! for Windows 95 was also available on floppy disks. DMF was a special 19-sector format Microsoft used to store 1.7MB on floppies rather than the usual 1.44MB. While the floppy edition of Windows was normally on 3.5" disks, a 5.25" version could be specially ordered as well.

Legacy

Many features that have become key components of the Microsoft Windows series, such as the Start menu and the taskbar, originated in Windows 95. Neil MacDonald, a Gartner analyst, said "If you look at Windows 95, it was a quantum leap in difference in technological capability and stability." Windows 95 was the first version of Windows to be a standard factory install on the average PC. Ina Fried of *CNET* said "By the time Windows 95 was finally ushered off the market in 2001, it had become a fixture on computer desktops around the world."[]

References

[4] Internet Archive

Further reading

Microsoft:

Third-party:

Windows_NT_4.0

Windows NT 4.0 is a preemptive,[1] graphical and business-oriented operating system designed to work with either uniprocessor or symmetric multi-processor computers. It was part of Microsoft's Windows NT line of operating systems and was released to manufacturing on 31 July 1996.[2] It is a 32-bit Windows system available in both workstation and server editions with a graphical environment similar to that of Windows 95.

Overview

The successor to Windows NT 3.51, Windows NT 4.0 introduced the modern user interface of Windows 95 to the Windows NT product line, including the Windows Shell, Windows Explorer (known as Windows NT Explorer), and the use of "My" nomenclature (e.g. My Computer). It also includes most applications introduced with Windows 95. Internally, Windows NT 4.0 was known as the Shell Update Release (SUR).[3] Various administrative tools, notably *User Manager for Domains*, *Server Manager* and *Domain Name Service Manager* have improved graphical user interfaces. The Start Menu in Windows NT 4.0 separated the *per-user* shortcuts and folders from the *All users* shortcuts and folders by a separator line.[4] Windows NT 4.0 includes some enhancements from Microsoft Plus! for Windows 95 such as the 3D Pinball game, font smoothing, full window drag, high color icons and stretching the wallpaper to fit the screen. Windows Desktop Update could also be installed on Windows NT 4.0 to update the shell version and install Task Scheduler.[5] The Windows NT 4.0 Resource Kit included the Desktop Themes utility.[6]

Windows NT 4.0 is the last major release of Microsoft Windows to support the Alpha, MIPS or PowerPC CPU architectures. It remained in use by businesses for a number of years, despite Microsoft's many efforts to get customers to upgrade to Windows 2000 and newer versions. It was also the last release in the Windows NT line to be branded as *Windows NT.*

Features

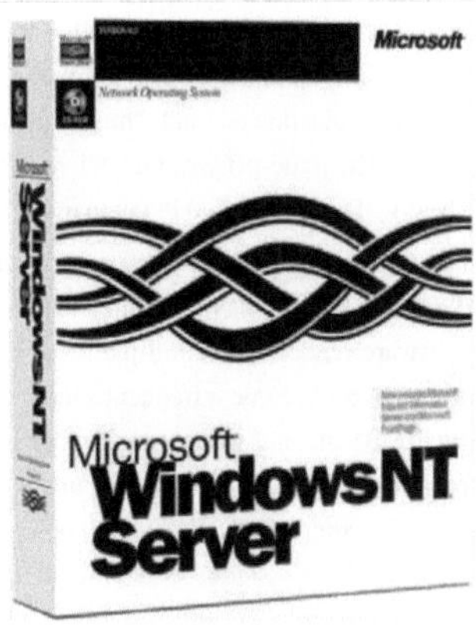

Windows NT 4.0 Server edition

Although the chief enhancement has been the addition of the Windows 95 shell, there are several major performance, scalability and feature improvements to the core architecture, kernel, USER32, COM and MSRPC.[3] [7] Windows NT 4.0 also introduced the concept of *System Policies* [8] and the System Policy Editor.

Other important features included with this release were the Crypto API,[3] Telephony API 2.0 with limited Unimodem support,[9] which was the first release of TAPI on Windows NT, DCOM and new OLE features,[10] and Microsoft Transaction Server for network applications, Microsoft Message Queuing (MSMQ), which improved interprocess communication, Winsock 2 and the TCP/IP stack improvements, and file system defragmentation support.[11]

The server editions of Windows NT 4.0 include Internet Information Services 2.0, Microsoft FrontPage 1.1, NetShow Services, Remote Access Service (which includes a PPTP server for VPN functionality) and Multi-Protocol Routing service. There are new administrative wizards and a lite version of the Network Monitor utility shipped with System Management Server. The Enterprise edition introduced Microsoft Cluster Server.

One significant difference from previous versions of Windows NT is that the Graphics Device Interface (GDI) is moved into kernel mode[12] rather than being in user mode in the CSRSS process. This eliminated a process-to-process context switch in calling GDI functions, resulting in a significant performance improvement over Windows NT 3.51, particularly in the graphical user interface. This however also mandated that graphics and printer drivers had to run in kernel mode as well,[13] resulting in potential stability issues.

Windows NT 4.0 was the first release of Microsoft Windows to include DirectX as standard—version 2 shipped with the initial release of Windows NT 4.0, and version 3 was included with the release of Service Pack 3 in mid-1997. Unlike Windows 95 (which did not include DirectX until the OSR2 release in August 1996), Windows NT 4.0 does not support Direct3D and USB. Later versions of DirectX were not released for Windows NT 4.0, although an unofficial DirectX 5 package was available. However, OpenGL hardware-accelerated graphics were solidly supported since the very first moment and successfully used by many video games and 3D applications (i. e. Quake I, II and III, Unreal, Unreal Tournament, 3D Studio MAX, SoftImage, Maya...).

In early releases of 4.0, numerous stability issues did occur as graphics and printer vendors had to change their drivers to be compatible with the kernel mode interfaces exported to them by GDI. The change to move the GDI to run in the same process context as its caller was prompted by complaints from NT Workstation users about realtime graphics performance, but this change put a considerable onus on hardware manufacturers to update device drivers. Even when manufacturers, primarily graphics hardware manufacturers, wrote 4.0-specific drivers, the BSOD was much more prevalent in Windows NT 4.0 until these driver issues were resolved.

Windows NT 4.0 also included a new Windows Task Manager application. Previous versions of Windows NT included the Task List application, but it only shows *applications* currently on the desktop. To monitor how much CPU and memory resources are being used, users were forced to use Performance Monitor. The task manager offers a more convenient way of getting a snapshot of all the processes running on the system at any given time.

Microsoft offered up to Internet Explorer 6.0 SP1 for Windows NT 4.0 with Service Pack 6. Sysprep was introduced as a deployment tool with Windows NT 4.0.

Windows NT 4.0 upgraded NTVDM's x86 emulation in the RISC versions from 286 to 486.[14]

Comparison with Windows 95

While providing much greater stability than Windows 95, it was also less flexible from a desktop perspective. Much of the stability was gained by the use of protected memory and the hardware abstraction layer. Direct hardware access was disallowed and "misbehaving" applications were terminated without needing the computer to be restarted. The trade-off was that NT required much more memory (32 MB for normal desktop use, 128 MB or more for heavy 3D applications) in comparison to consumer targeted products such as Windows 95.[15]

While nearly all programs written for Windows 95 will run on Windows NT, many 3D games would not, due in part to Windows NT 4.0 having limited support for DirectX (however, it was possible to play many of them in OpenGL or software rendering). Third-party device drivers were an alternative to access the hardware directly, but poorly written drivers became a frequent source of "stop errors". Such failures began to be referred to as the "blue screen of death" or BSOD and would require the system to be restarted in such cases. These errors were very rare if using the appropriate drivers and it was not uncommon for NT servers or workstations to run for months at a time without failure. By comparison, Windows consumer versions at the time were much less stable and popularized the belief that all Windows versions were unreliable..

Windows NT 4.0 is also less user-friendly than Windows 95 when it comes to certain maintenance and management tasks; for instance, in spite of shipping a year later than Windows 95, by default there is no Plug and Play support and no Device Manager (although limited support could be installed later) which greatly simplifies installation of hardware devices. Many basic DOS applications would run, however graphical DOS applications would not run due to the way they accessed graphics hardware. Although Windows NT 4.0 introduced APIs for defragmentation,[11] there was no built-in defrag utility, unlike Windows 95. Also, Windows NT 4.0 lacked USB support, a preliminary version of which would be added to OEM editions of Windows 95 in OSR 2.1.

The difference between the NT/2000 and 95/98 lines of Windows ended with the arrival of the different versions of Windows XP. At that time, the APIs —such as OpenGL and DirectX— had matured sufficiently to be more efficient to write for common PC hardware. On the other hand, the hardware itself had become powerful enough to handle the API processing overhead acceptably.

The maximum amount of supported physical RAM in Windows NT 4.0 supports is 4 GB.[16] Windows 95 fails to boot on computers with approximately more than 480 MB of memory.[17]

Editions

Windows NT 4.0 Server was included in versions 4.0 and 4.5 of BackOffice Small Business Server suite.

Windows NT 4.0 Workstation edition

Client

- **Windows NT 4.0 Workstation** was designed for use as the general business desktop operating system.

Servers

- **Windows NT 4.0 Server**, released in 1996, was designed for small-scale business server systems.
- **Windows NT 4.0 Server, Enterprise Edition**, released in 1997, is the precursor to the *Enterprise* line of the Windows server family (*Advanced Server* in Windows 2000). Enterprise Server was designed for high-demand, high-traffic networks. Windows NT 4.0 Server, Enterprise Edition includes Service Pack 3.[18]
- **Windows NT 4.0 Terminal Server Edition**, released in 1998, allows the users to log on remotely. The same functionality was called *Terminal Services* in Windows 2000 and later server releases, and also powers the *Remote Desktop* feature that first appeared in Windows XP.

Embedded

- **Windows NT 4.0 Embedded** (abbreviated **NTe**) is an edition of Windows NT 4.0 that was aimed at computer-powered major appliances, vending machines, ATMs and other devices that cannot be considered general-purpose computers *per se*. It is the same system as the standard Windows NT 4.0, but it comes packaged in a database of components and dependencies, from which a developer can choose individual components to build customized setup CDs and hard disk boot images. Windows NT 4.0 Embedded includes Service Pack 5. It was succeeded by Windows XP Embedded.

Upgradeability

An "option pack" was available as a free-bundled CD starting around 1998, which included IIS 4.0 with Active Server Pages, FrontPage Server Extensions, Certificate Server, MTS, MSMQ, CDONTS, Internet Authentication Service (IAS), Indexing Service, Microsoft Management Console 1.0, Microsoft Site Server, SMTP and NNTP services and other new software.

Several features such as Distributed File System and Windows NT Load Balancing Service (WLBS) were delivered as addons for Windows NT Server 4.0. The Routing and Remote Access Service was also a downloadable feature which replaced Windows NT 4.0's separate RAS and Multi-Protocol Routing services.

The last version of Microsoft Office to run on Windows NT 4.0 is Office XP.

Service packs

Service pack	Release date
Service Pack 1 (SP1)	16 October 1996
Service Pack 2 (SP2)	14 December 1996
Service Pack 3 (SP3)	15 May 1997
Service Pack 4 (SP4)	25 October 1998
Service Pack 5 (SP5)	4 May 1999
Service Pack 6 (SP6)	22 November 1999
Service Pack 6a (SP6a)	30 November 1999

Microsoft released Windows NT 4.0 service packs primarily to fix bugs. Windows NT 4.0 , during the product's lifecycle, had several service packs, as well as numerous service rollup packages and option packs. The last full service pack was Service Pack 6a (SP6a).

A SP7 was planned at one stage in early 2001, but this became the *Post SP6a Security Rollup* and not a full service pack, released on 26 July 2001, 16 months after Windows 2000 and nearly three months prior to Windows XP.[19]

The service packs also added a multitude of new features such as newer versions of or improvements to Internet Information Services, public-key and certificate authority functionality, user accounts and user profile improvements, smart card support, improved symmetric multiprocessing (SMP) scalability, clustering capabilities, COM support improvements, User Profile Disk Quotas, Event Log service, *Security Configuration Manager* MMC snap-in, MS-CHAPv2 and NTLMv2, SMB packet signng, SYSKEY, boot improvements, WINS improvements, Routing and Remote Access Service (RRAS), PPTP, DCOM/HTTP tunneling improvements, IGMPv2, WMI, Active Accessibility and NTFS 3.0 support among others.[20]

Resource Kits

Microsoft released five revisions of the Windows NT 4.0 Workstation and Server Resource Kit (original release plus four supplements) which contained a large number of tools and utilities as well as third-party software.

Security

Microsoft stopped providing security updates for Windows NT 4.0 Workstation on 30 June 2004 and Windows NT 4.0 Server on 31 December 2004, due to major security flaws including Microsoft Security Bulletin MS03-010 [21], which according to Microsoft could not be patched without significant changes to the core operating system. According to the security bulletin, "Due to [the] fundamental differences between Windows NT 4.0 and Windows 2000 and its successors, it is infeasible to rebuild the software for Windows NT 4.0 to eliminate the vulnerability. To do so would require re-architecting a very significant amount of the Windows NT 4.0 operating system, and [...] there would be no assurance that applications designed to run on Windows NT 4.0 would continue to operate on the patched system."

Between June 2003 and June 2007, 127 security flaws were identified and patched in Windows 2000 Server, many of which may also affect Windows NT 4.0 Server; however, Microsoft does not test security bulletins against unsupported software.

Legacy

The stability of Windows NT offered reduced support costs over Windows 95 or Windows 98. It was later succeeded by Windows 2000 which was based on NT and largely bridged the gap between NT and consumer Windows versions. Windows XP and later versions were released which completed the unification of the core architecture of all currently marketed Windows versions around NT.

References

[2] Microsoft Announces the Release of Windows NT Workstation 4.0 (31 July 1996) http://www.microsoft.com/presspass/press/1996/jul96/ntw40pr.mspx

[3] Poking Around Under the Hood: A Programmer's View of Windows NT 4.0 (http://www.microsoft.com/msj/archive/S413.aspx)

[4] Windows 2000 Professional Beta 3 Review (http://www.winsupersite.com/article/windows-2000/windows-2000-professional-beta-3-reviewed-127366)

[5] The New Task Scheduler (Windows 95 and Windows NT 4.0) (http://technet.microsoft.com/en-us/library/cc751232.aspx)

[6] Windows NT 4.0 Resource Kit Utilities Corrections and Comments (http://support.microsoft.com/kb/159564)

[7] Microsoft, DCE, and COM (ftp://ftp.microsoft.com/Developr/drg/COM/MarkRy/MS_DCE_COM.ppt)

[8] Guide To Windows NT 4.0 Profiles and Policies (http://support.microsoft.com/kb/161334)

[9] For the Telephony API, Press 1; For Unimodem, Press 2; or Stay on the Line: Microsoft Systems Journal article (http://www.microsoft.com/msj/0498/tapi.aspx)

[10] Introducing Distributed COM and the New OLE Features in Windows NT 4.0 (http://www.microsoft.com/msj/0596/activex0596.aspx)

[11] Inside Windows NT Disk Defragmenting (http://bandwidthco.com/whitepapers/compforensics/fsanalysis/fragmentation/Inside Windows NT Disk Defragmenting.pdf)

[13] Converting Win32 Kernel-mode Print Drivers to User Mode (http://www.unixwiz.net/techtips/win32-pdriver-ktou.html)

[14] INFO: How Windows handles floating-point calculations (http://support.microsoft.com/kb/102555)

[18] Windows NT Server 4.0, Enterprise Edition, Administrator's Guide and Release Notes (http://technet.microsoft.com/en-us/library/cc767919.aspx)

[20] What's New in Windows NT 4.0 Service Pack 4? (http://web.archive.org/web/19990117055557/http://www.microsoft.com/ntserver/nts/exec/overview/NT4SP4whatnew.asp)

[21] http://www.microsoft.com/technet/security/Bulletin/MS03-010.mspx

External links

- Guidebook: Windows NT 4.0 Gallery (http://www.guidebookgallery.org/guis/windows/winnt40) – A website dedicated to preserving and showcasing Graphical User Interfaces
- HPC:Factor Windows NT 4.0 Workstation Patches & Updates Guide (http://www.hpcfactor.com/qlink/?linkID=16)
- HPC:Factor Windows NT 4.0 Server Patches & Updates Guide (http://www.hpcfactor.com/qlink/?linkID=149)
- Josephn.net: Windows NT 4.0 Terminal Server Edition Tips & Updates (http://www.josephn.net/windows_nt4_resources)
- MDGx: Windows NT 4.0 Essential Free Upgrades + Fixes (http://www.mdgx.com/wnt4.htm)

Windows_Explorer

File Explorer, previously known as **Windows Explorer**, is a file manager application and also a navigation tool that is included with releases of the Microsoft Windows operating system from Windows 95 onwards. It provides a graphical user interface for accessing the file systems. It is also the component of the operating system that presents many user interface items on the monitor such as the taskbar and desktop. Controlling the computer is possible without Windows Explorer running (for example, the File | Run command in Task Manager on NT-derived versions of Windows will function without it, as will commands typed in a command prompt window). Located in the C:\Windows directory, it is sometimes referred to as the Windows shell, explorer.exe, or simply "Explorer".

Overview

What was Windows Explorer was first included with Windows 95 as a replacement for File Manager, which came with all versions of Windows 3.x operating systems. Explorer could be accessed by double-clicking the new *My Computer* desktop icon, or launched from the new *Start Menu* that replaced the earlier Program Manager. There is also a shortcut key combination: Windows key + E. Successive versions of Windows (and in some cases, Internet Explorer) introduced new features and capabilities, removed other features, and generally progressed from being a simple file system navigation tool into a task-based file management system.

While "Windows Explorer" or "File Explorer" is a term most commonly used to describe the file management aspect of the operating system, the Explorer *process* also houses the operating system's search functionality and File Type associations (based on filename extensions), and is responsible for displaying the desktop icons, the Start Menu, the Taskbar, and the Control Panel. Collectively, these features are known as the Windows shell.

After a user logs in, the explorer process is created by userinit process. Userinit performs some initialization of the user environment (such as running the login script and applying group policies) and then looks in the registry at the Shell value and creates a process to run the system-defined shell – by default, Explorer.exe. Then Userinit exits. This is why Explorer.exe is shown by various process explorers with no parent – its parent has exited.

History

In 1995, Microsoft first released test versions of a shell refresh, named the *Shell Technology Preview*, and often referred to informally as "NewShell".[1] The update was designed to replace the Windows 3.x Program Manager/File Manager based shell with Windows Explorer. The release provided capabilities quite similar to that of the Windows "Chicago" (codename for Windows 95) shell during its late beta phases, however was intended to be nothing more than a test release.[2] There were two public releases of the Shell Technology Preview, made available to MSDN and CompuServe users: 26 May 1995 and 8 August 1995. Both held Windows Explorer builds of 3.51.1053.1. The Shell Technology Preview program never saw a final release under NT 3.51. The entire program was moved across to the Cairo development group who finally integrated the new shell design into the NT code with the release of NT 4.0 in July 1996.

Windows 98 and Windows Desktop Update

With the release of the Windows Desktop Update (packaged with Internet Explorer 4.0 as an optional component, and included in Windows 98), Windows Explorer became "integrated" with Internet Explorer, most notably with the addition of navigation arrows (back and forward) for moving between recently visited directories, as well as Internet Explorer's Favorites menu. At the time these changes raised antitrust concerns about the incorporation of what was seen as an application feature, but this feature has since been emulated by most other file browsers.

An address bar was also added to Windows Explorer, which a user could type in directory paths directly, and be taken to that folder. It also operated as a URL bar for Internet addresses; web pages would open in the main portion of the window.

Another feature that was based on Internet Explorer technology was customized folders. Such folders contained a hidden web page that controlled the way the Windows Explorer displayed the contents of the folder. This feature proved to have security vulnerabilities due to its reliance on ActiveX objects and scripting, and was removed with the introduction of Windows XP.

Other new features:

- The ability to add other toolbars to the task bar, the most visible of these being Quick Launch.
- Active Desktop, which made it possible to turn the desktop background itself into a web page.
- Single-click activation of icons in Windows Explorer, adhering to a web page paradigm.
- Desktop Channels.
- New virtual folders for Scheduled Tasks and Web Folders

Windows Me and Windows 2000

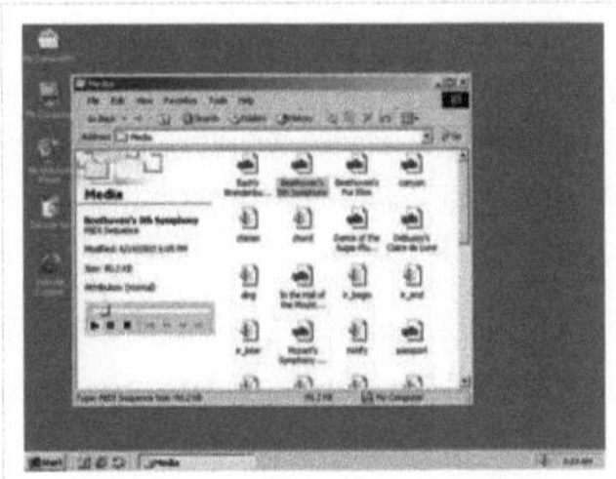

The integrated media player in Windows Explorer playing a MIDI sequence.

The "Web-style" folders view, with the left Explorer pane displaying details for the object currently selected, is turned on by default. For certain file types, such as pictures and media files, a preview is also displayed in the left pane.[3] The Windows 2000 Explorer featured an interactive media player as the previewer for sound and video files. However, such a previewer can be enabled in Windows Me through the use of folder customization templates. [4] Windows Explorer in Windows 2000 and Windows Me allows for custom thumbnail previewers and tooltip handlers. The default file tooltip displays file title, author, subject and comments;[] this metadata may be read from a special NTFS stream, if the file is on an NTFS volume, or from an COM Structured Storage stream, if the file is a structured storage document. All Microsoft Office documents since Office 95[5] make use of structured storage, so their metadata is displayable in the Windows 2000 Explorer default tooltip. File shortcuts can also store comments which are displayed as a tooltip when the mouse hovers over the shortcut.

The right-hand pane, which usually just lists files and folders, can also be customized. For example, the contents of the system folders aren't displayed by default, instead showing in the right pane a warning to the user that modifying the contents of the system folders could harm their computer. It's possible to define additional Explorer panes by using DIV elements in folder template files.[] This feature was abused by computer viruses that employed malicious scripts, Java applets, or ActiveX controls in folder template files as their infection vector. Two such viruses are VBS/Roor-C[6] and VBS.Redlof.a.[]

Other Explorer UI elements that can be customized include columns in "Details" view, icon overlays, and search providers: the new DHTML-based search pane is integrated into Windows 2000 Explorer, unlike the separate search dialog found in all previous Explorer versions.[]

Search capabilities were added, offering full-text searches of documents, with options to filter by date (including arbitrary ranges like “modified within the last week”), size, and file type. The Indexing Service has also been integrated into the operating system and the search pane built into Explorer allows searching files indexed by its database.[7] The ability to customize the standard buttons was also added.

Windows XP and Windows Server 2003

There were significant changes made to Windows Explorer in Windows XP, both visually and functionally. Microsoft focused especially on making Explorer more discoverable and task-based, as well as adding a number of features to reflect the growing use of a computer as a "digital hub".

Windows Explorer in Windows Server 2003 contains all the same features as Windows XP, but the task panes and search companion are disabled by default.

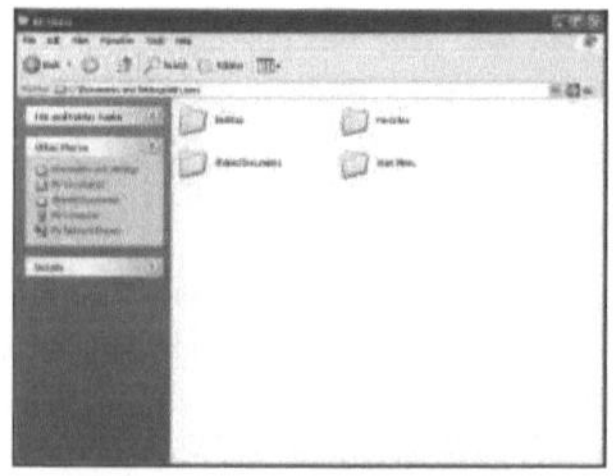

Windows Explorer in Windows XP

Task Pane

The task pane is displayed on the left-hand side of the window instead of the traditional folder tree view. It presents the user with a list of common actions and destinations that are relevant to the current directory or file(s) selected. For instance, when in a directory containing mostly pictures, a set of "Picture tasks" is shown, offering the options to display these pictures as a slide show, to print them out, or to go online to order prints. Conversely, a folder containing music files would offer options to play those files in a media player, or to go online to purchase music. Windows XP had a Media bar but it was removed with SP1. The Media Bar was only available with Windows XP RTM.

Every folder also has "File and Folder Tasks", offering options to create new folders, share a folder on the local network, publish files or folders to a web site, and other common tasks like copying, renaming, moving, and deleting files or folders. File types that have identified themselves as being printable also have an option listed to print the file.

Underneath "File and Folder Tasks" is "Other Places", with links to other common locations such as "My Computer", "Control Panel", and "My Documents". These also change depending on what folder the user was in, leading to some criticism of Microsoft for not being consistent in displaying navigation choices.

Underneath "Other Places" is a "Details" pane which gives additional information – typically file size and date, but depending on the file type, a thumbnail preview, author, image dimensions, or other details.

The "Folders" button on the Windows Explorer toolbar toggles between the traditional tree view of folders, and the task pane. Users can get rid of the task pane or restore it using the sequence: Tools – Folder Options – General – Show Common Tasks/Use Windows Classic Folders.

Search companion

Microsoft introduced animated "Search Companions" in an attempt to make searching more engaging and friendly; the default character is a puppy named Rover (previously used in Microsoft Bob), with three other characters (Merlin the magician, Earl the surfer, and Courtney) also available. These search companions bear a great deal of similarity to Microsoft Office's Office Assistants, even incorporating "tricks" and sound effects.

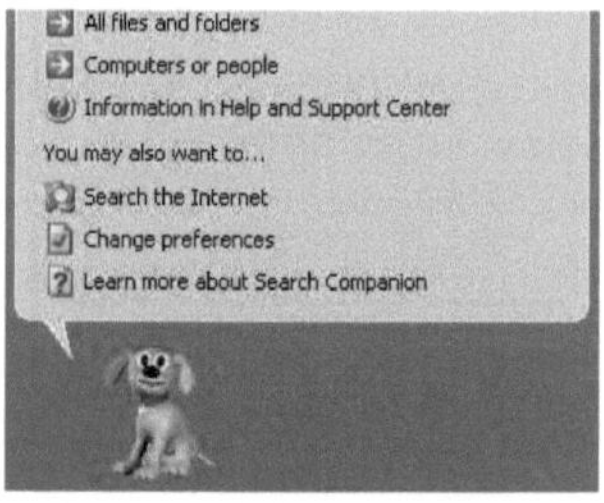

Windows Explorer's default Search Companion, Rover.

The search capability itself is fairly similar to Windows Me and Windows 2000, with one major addition: Search can also be instructed to search only files that are categorically "Documents" or "Pictures, music and video"; this feature is noteworthy largely because of how

Windows determines what types of files can be classified under these categories. In order to maintain a relevant list of file types, Windows Explorer connects to Microsoft and downloads a set of XML files that define what these file types are. While harmless in itself, this feature caught the attention of a number of privacy advocates, and users with firewall software installed, who felt it was unnecessary for Windows to connect to Microsoft's servers whenever a local search was performed.

Image handling

Windows XP improves image preview in Explorer by offering a Filmstrip view. "Back" and "Previous" buttons facilitate navigation through the pictures, and a pair of "Rotate" buttons offer 90-degree clockwise and counter-clockwise (lossy) rotation of images. Aside from the Filmstrip view mode, there is a 'Thumbnails' mode, which displays thumbnail-sized images in the folder. A Folder containing images will also show thumbnails of four of the images from that folder overlaid on top of a large folder icon.

Folder thumbnail preview

Web publishing

Web sites that offer image hosting services can be plugged into Windows Explorer, which the user can use to select images on their computer, and have them uploaded correctly without dealing with comparatively complex solutions involving FTP or web interfaces.

Other changes

- Explorer gained the ability to understand the metadata of a number of types of files. For example, with images from a digital camera, the Exif information can be viewed, both in the Properties pages for the photo itself, as well as via optional additional Details View columns.
- A Tile view mode was added, which displays the file's icon in a larger size (48 × 48), and places the file name, descriptive type, and additional information (typically the file size for data files, and the publisher name for applications) to the right.
- The Details view also presented an additional option called "Show in Groups" which allows the Explorer to separate its contents by headings based on the field which is used to sort the items.
- The taskbar can be locked to prevent it from accidentally being moved.
- Windows Explorer also gained the ability to burn CDs and DVD-RAM discs in Windows XP.

Removed features

- The sort order has changed compared to the one in Windows 2000. For file names containing numbers Windows Explorer now tries to sort based on numerical value rather than just comparing each number digit by digit for every character position in the file name.[]

Windows Vista and Windows Server 2008

Search, organizing & metadata

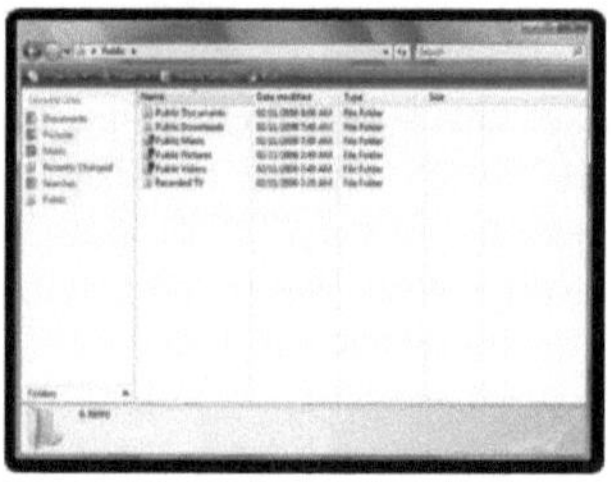

Windows Explorer in Windows Vista

Windows Explorer includes significant changes from previous versions of Windows such as improved filtering, sorting, grouping and stacking. Combined with integrated desktop search, Windows Explorer allows users to find and organize their files in new ways, such as *Stacks*. The "Stacks" view groups files according to the criterion specified by the user. Stacks can be clicked to filter the files shown in Windows Explorer. There is also the ability to save searches as virtual folders or *Search Folders*. A *Search Folder* is simply an XML file, which stores the query in a form that can be used by the Windows search subsystem. When accessed, the search is executed, and the results are aggregated and presented as a virtual folder. When sorting items, the sort order no longer remains consistently Ascending or Descending. Each property has a preferred sort direction. For example, *Sort by Date* defaults to Descending order, as does *Size*. But *Name* and *Type* default to Ascending order.

Searching for files containing a given text string became problematic with Vista unless the files have been indexed. An alternative is to use the findstr command-line function.[8] After right-clicking on a folder one can open a command-line prompt in that folder.

Windows Explorer also contains modifications in the visualization of files on a computer. A new addition to Windows Explorer in Vista and Server 2008 is the *Details pane*, which displays metadata and information relating to the currently selected file or folder. The *Details pane* will also display a thumbnail of the file or an icon of the filetype if the file does not contain visual information. Furthermore, different imagery is overlayed on thumbnails to give more information about the file, such as a picture frame around the thumbnail of an image file, or a filmstrip on a video file. Thumbnails can be zoomed on.

The details pane in Windows Vista and Windows Server 2008 showing metadata of an image.

The *Details pane* also allows for the change of some textual metadata such as 'Author' and 'Title' in files that support them within Windows Explorer. A new type of metadata called *tags* allows users to add descriptive terms to documents for easier categorization and retrieval. Some files support open metadata, allowing users to define new types of metadata for their files. Out-of-the-box, Windows Vista and Windows Server 2008 supports Microsoft Office documents and most audio and video files. Support for other file types can however be added by writing specialized software to retrieve the metadata at the shell's request. Metadata stored in a file's alternate (secondary) stream only on NTFS volumes cannot be viewed and edited through the 'Summary' tab of the file's properties anymore. Instead, all metadata is stored inside the file, so that it will always travel with the file and not be dependent on the file system.[9]

Layout and icons

Windows Explorer in Windows Vista and Windows Server 2008 also introduces a new layout. The Task Panes from Windows XP are replaced with a toolbar on top and a *Navigation pane* on the left. The *Navigation pane* contains commonly accessed folders and pre-populated *Search folders*. Seven different views are available to view files and folders, namely, *List, Details, Small icons, Medium icons, Large icons, Extra large icons or Tiles*. *Automatic folder type discovery* automatically detects the contents of a folder and display the correct detail settings. File and folder actions such as Cut, Copy, Paste, Undo, Redo, Delete, Rename and Properties are built into a dropdown menu which appears when the *Organize* button is clicked. It is also possible to change the layout of the Explorer window by using the Organize button. Users can select whether to display *Classic Menus*, a *Search Pane*, a *Preview Pane*, a *Reading Pane*, and/or the *Navigation Pane*. The *Preview pane*, allows for previewing files (viewing documents and playing media files) in addition to the *Details pane*. Also, Explorer can show a preview for any image format if the necessary codec for the format is installed in the *Windows Imaging Component*.

The address bar has been replaced by a *breadcrumbs bar* for easier navigation. It shows the full path to the current location. Clicking any location in the path hierarchy takes the user to that level, instead of repeatedly pressing the Back button. This is similar to what is possible in Windows XP by pressing the small down-arrow next to "Back" and selecting any folder from a list of previously accessed folders. It is also possible to navigate to any subfolder of the current folder using the arrow to the right of the last item, or to click in the space to the right of this to copy or edit the path manually. As with many other Microsoft-made Windows Vista applications, the menu bar is hidden by default. Pressing the Alt key makes the menu bar appear.

Check boxes in Windows Explorer allow the selection of multiple files. Free and used space on all drives is shown in horizontal indicator bars. Icons of various sizes are supported – *16 x 16, 24 x 24, 32 x 32, 48 x 48, 64 x 64, 96 x 96, 128 x 128 and 256 x 256*. Windows Explorer can zoom the icons in and out using a slider or by holding down the *Ctrl* key and using the mouse scrollwheel.

Other changes

With the release of Windows Vista & Server 2008 and Windows Internet Explorer 7 for Windows XP, Internet Explorer is no longer integrated with Windows Explorer. In Windows Vista and Server 2008 (and in Windows XP as well if IE7 or 8 is installed), Windows Explorer no longer displays web pages, and IE7 does not support use as a file manager, although one will separately launch the other as necessary.

When moving or copying files from one folder to another, if two files have the same name, an option is now available to rename the file; in previous versions of Windows, the user was prompted to choose either a replacement or cancel moving the file. Also, when renaming a file, Explorer only highlights the filename without selecting the extension. Renaming multiple files is quicker as pressing Tab automatically renames the existing file or folder and opens the file name text field for the next file for renaming. Shift+Tab allow renaming in the same manner upwards.

Support for burning data on DVDs (DVD±R, DVD±R DL, DVD±R RW) in addition to CDs and DVD-RAM using version 2.0 of the Image Mastering API was added.

In case a file is in use by another application, Windows Explorer informs users to close the application and retry the file operation. Also, a new interface 'IFileIsInUse' is introduced into the API which developers can use to let other applications switch to the main window of the application that has the file open or simply close the file from the "File In Use" dialog.[10] If the running application exposes these operations by means of the *IFileInUse* interface, Windows Explorer, upon encountering a locked file, allows the user to close the file or switch to the application from the dialog box itself.

Removed and changed features

The ability to customize the layout and buttons on the toolbars has been removed in Windows Vista's Explorer, as has the ability to add a password to a zip file (compressed folder). The Toolbar button in Explorer to go up one folder from the current folder has been removed (the function still exists however, one can move up a folder by pressing Alt + ↑). Although still fully available from the menus and keyboard shortcuts, toolbar buttons for Cut, Copy, Paste, Undo, Delete, Properties and some others are no longer available. The Menu Bar is also hidden by default, but is still available by pressing the *Alt* key, or changing its visibility in the layout options. Several other features are removed such as showing the size on the status bar without selecting items, storing metadata in NTFS secondary streams,[11] the IColumnProvider interface which allowed addition of custom columns to Explorer[12] and folder background customization using *desktop.ini.*

The option "Managing pairs of Web pages and folders" is also removed, and the user has no way of telling Vista that a .html file and the folder with the same name that was created when saving a complete web page from IE should be treated separately, that is, they cannot delete the folder without deleting the html file as well. A solution to this is provided on (http://windowsxp.mvps.org/webpairs.htm). After the webpairs.reg file has been merged into the registry, the "Managing pairs of Web pages and folders" option is available in the Folder Options View tab.

The ability to right-click a folder and hit "Search" was removed in Windows Vista Service Pack 1. Users must open the folder they wish to search in and enter their keywords in the search field located on the top right corner of the window. Alternatively, users can specify other search parameters through the "Advanced Search" UI, which can be accessed by clicking on the Organize Bar, and selecting Search Pane under the Layout submenu. Pressing F3 also opens the "Advanced Search" interface.

Windows 7 and Windows Server 2008 R2

Libraries

Windows Explorer in Windows 7

Windows Explorer in Windows 7 and Windows Server 2008 R2 supports *Libraries*, virtual folders described in a .library-ms file that aggregates content from various locations – including shared folders on networked systems if the shared folder has been indexed by the host system – and present them in a unified view. Searching in a library automatically federates the query to the remote systems, in addition to searching on the local system, so that files on the remote systems are also searched. Unlike search folders, Libraries are backed by a physical location which allows files to be saved in the Libraries. Such files are transparently saved in the backing physical folder. The default save location for a library may be configured by the user, as can the default view layout for each library. Libraries are generally stored in the *Libraries* special folder, which allows them to be displayed on the navigation pane.

By default, a new user account in Windows 7 contains four libraries, for different file types: Documents, Music, Pictures, and Videos. They are configured to include the user's profile folders for these respective file types, as well as the computer's corresponding Public folders.

In addition to aggregating multiple storage locations, Libraries enable Arrangement Views and Search Filter Suggestions. Arrangement Views allow users to pivot their views of the library's contents based on metadata. For example, selecting the "By Month" view in the Pictures library will display photos in stacks, where each stack represents a month of photos based on the date they were taken. In the Music library, the "By Artist" view will display stacks of albums from the artists in their collections, and browsing into an artist stack will then display the relevant albums.

Search Filter Suggestions are a new feature of the Windows 7 and Windows Server 2008 R2 Explorer's search box. When the user clicks in the search box, a menu shows up below it showing recent searches as well as suggested Advanced Query Syntax filters that the user can type. When one is selected (or typed in manually), the menu will update to show the possible values to filter by for that property, and this list is based on the current location and other parts of the query already typed. For example, selecting the "tags" filter or typing "tags:" into the search box will display the list of possible tag values which will return search results.

Arrangement Views and Search Filter Suggestions are database-backed features which require that all locations in the Library be indexed by the Windows Search service. Local disk locations must be indexed by the local indexer, and Windows Explorer will automatically add locations to the indexing scope when they are included in a library. Remote locations can be indexed by the indexer on another Windows 7 and Windows Server 2008 R2 machine, on a Windows machine running Windows Search 4 (such as Windows Vista or Windows Home Server), or on another device that implements the MS-WSP remote query protocol.[13]

Federated search

Windows Explorer also supports federating search to external data sources, such as custom databases or web services, that are exposed over the web and described via an OpenSearch definition. The federated location description (called a *Search Connector*) is provided as a .osdx file. Once installed, the data source becomes queryable directly from Windows Explorer. Windows Explorer features, such as previews and thumbnails, work with the results of a federated search as well.

Other changes

- Windows 7 and Windows Server 2008 R2 supports showing icons in the context menu and creating cascaded context menus with static verbs in submenus using the Registry instead of a shell extension.[14]
- The search box in the Explorer window and the address bar can be resized.
- Certain folders in the navigation pane can be hidden to reduce clutter.
- Progress bars and overlay icons on an application's button on the taskbar.
- Content view which shows thumbnails and metadata.
- Buttons to toggle the preview pane and create a new folder.

Removed and changed features

In Windows 7, several features have been removed from Windows Explorer, including the collapsible folder pane, overlay icon for shared items, remembering individual folder window sizes and positions, free disk space on the status bar, icons on the command bar, ability to disable Auto Arrange and Align to Grid, sortable column headings in other views except details view, ability to disable full row selection in details view, automatic horizontal scrolling and scrollbar in the navigation pane and maintaining selection when sorting from the Edit menu.

Windows 8 and Windows Server 2012

The new File Explorer interface in Windows 8

The file manager on Windows 8 and Windows Server 2012 is renamed **File Explorer**, and introduces new features such as a redesigned interface incorporating a ribbon toolbar, and a redesigned file operation dialog that displays more detailed progress and allows for file operations to be paused and resumed.[15] []

Extensibility

File Explorer can be extended to support non-default functionality by means of *Windows Shell Extensions*, which are COM objects that plug the extended functionality into Windows Explorer.[16] Shell extensions can be in the form of shell extension handlers, toolbars or even namespace extensions that allow certain folders (or even non-filesystem objects such as the images scanned by a scanner) to be presented as a special folder. File Explorer also allows metadata for files to be added as NTFS Alternate Data Streams, separate from the data stream for the file.

Shell extension handlers are queried by the shell beforehand for modifying the action the shell takes. They can be associated on a per-file type basis – where they will show up only when a particular action takes place on a particular file type – or on a global basis – which are always available. The shell supports the following extension handlers:

Handler	Description	Can be implemented on	Required shell version
Context menu handler	Adds menu items to the context menu. It is called before the context menu is displayed.	Per-file type basis	Windows 95 and later. Windows 7 introduced IExecuteCommand
Drag-and-drop handler	Controls the action upon right-click drag and drop and modifies the context menu that appears.	Global basis	Windows 95 and later
Drop target handler	Controls the action after a data object is dragged and dropped over a drop target such as a file.	Per-file type basis	Windows 95 and later
Data object handler	Controls the action after a file is copied to the clipboard or dragged and dropped over a drop target. It can provide additional clipboard formats to the drop target.	Per-file type basis	Windows 95 and later
Icon handler	Assigns a custom icon to an individual file amongst a class of file types. It is called before file icons are displayed.	Per-file type basis	Windows 95 and later
Property sheet handler	Replaces or adds pages to the property sheet dialog box of an object.	Per-file type basis	Windows 95 and later
Copy hook handler	Allows running, modifying or denying the action when a user or application tries to copy, move, delete, or rename an object.	*Not associated with a file type*	Windows 95 and later
Search handler	Allows shell integration of a custom search engine.	*Not associated with a file type*	Windows 95 and later up to Windows XP
Infotip handler	Allows retrieving flags and infotip information for an item and displaying it inside a popup tooltip upon mouse hover.	Per-file type basis	Windows Desktop Update and later
Thumbnail image handler	Provides for a thumbnail image to be generated and displayed along with its alpha type when a file is selected or the thumbnail view is activated.	Per-file type basis	Windows Desktop Update and later. Windows Vista introduced a newer IThumbnailProvider interface that also shows thumbnails in the Details pane. The older IExtractImage is still supported but not in the Details pane.[17]
Disk Cleanup handler	Add a new entry to the Disk Cleanup application and allows specifying additional disk locations or files to clean up.	Per-folder basis	Windows 98 and later
Column handler	Allows creating and displaying custom columns in Windows Explorer *Details view*. It can be used to extend sorting and grouping.	Per-folder basis	Windows 2000, Windows Me, Windows XP & Server 2003

Icon overlay handler	Allows displaying an overlay icon over a shell object (a file or folder icon).	Per-file type basis	Windows 2000 and later
Metadata handler	Allows viewing and modifying metadata stored in a file. It can be used to extend details view columns, infotips, property pages, sorting and grouping.	Per-file type basis	Windows 2000 and later
Filter handler (IFilter)	Allows file properties and its contents to be indexed and searched by Indexing Service or Windows Search	Per-file type basis	Windows 2000 and later
AutoPlay handler	Examines newly discovered removable media and devices and, based on content such as pictures, music or video files, launches an appropriate application to play or display the content.	Per file type category. In Windows XP only, per-device and per file type category.	Windows XP and later
Property handler	Allows viewing and modifying system-defined and custom properties of a file.	Per-file type basis	Windows Vista and later, on Windows XP if Windows Search is installed.
Preview handler	Renders enhanced previews of items without launching the default application when a file is selected. It can also provide file type-specific navigation such as browsing a document, or seeking inside a media file.	Per-file type basis	Windows Vista and later

Namespace extensions are used by Explorer to either display some data – which are not persisted as files – in a folder-like view or to present data in a way that is different from their organization on the file system. This feature can be exploited by a relational file system like liquidFOLDERs or Tabbles, clones of the ill-fated Microsoft WinFS. Special Folders, such as *My Computer* and *Network Places* in Windows Explorer are implemented this way, as are Explorer views that let items in a mobile phone or digital camera be explored. Source-control systems that use Explorer to browse source repositories also use Namespace extensions to allow Explorer to browse the revisions. To implement a namespace extension, the IPersistFolder, IShellView, IShellFolder, IShellBrowser and IOleWindow interfaces needs to be implemented and registered. The implementation needs to provide the logic for navigating the data store as well as describing the presentation. Windows Explorer will instantiate the COM objects as required.[]

While Windows Explorer natively exposes the extensibility points as COM interfaces, .NET Framework can also be used to write extensions, using the COM Interop functionality of .NET Framework.[] While Microsoft itself makes available extensions – such as the *Photo Info tool*[18] – which are authored using .NET Framework, they currently recommend against writing managed shell extensions, as only one instance of the CLR (prior to version 4.0) can be loaded per-process. This behavior would cause conflicts if multiple managed add-ins, targeting different versions of the CLR, are attempted to be run simultaneously.[19] [20]

Limitations

Even in Windows 7, Explorer uses the ANSI API calls, so it is limited to operating on paths that have fewer than 260 (MAX_PATH in windef.h) characters. The limitation applies to some operations like copying, moving or deleting files (to the recycle bin), but not to simply accessing files by double-click. Other Microsoft file tools, like Robocopy, do not have this limitation.

See also

- Comparison of file managers
- File manager
- Windows shell replacement

Notes and references

[1] Window NT 3.51 tricks (http://toastytech.com/guis/misc4.html)
[2] Internet Archive
[3] Managing Files, Folders, and Search Methods: Microsoft TechNet (http://www.microsoft.com/technet/prodtechnol/windows2000serv/reskit/prork/prde_ffs_sleq.mspx?mfr=true)
[4] WebView for WinMe (http://www.mvps.org/serenitymacros/home.html)
[6] *Sophos*, VBS/Roor-C threat analysis (http://www.sophos.com/security/analyses/vbsroorc.html). Accessed 2007-08-26.
[7] What is Indexing Service?: MSDN (http://msdn.microsoft.com/en-us/library/ms689718.aspx)
[8] (http://www.wikihow.com/Search-for-Contents-in-Any-File-Type-Without-Indexing-Service-Enabled-in-Windows-Vista-and-Windows-7)
[9] Windows Vista Help: Add tags or other properties to files (http://windowshelp.microsoft.com/Windows/en-US/help/0acda837-dacc-4a88-bb01-43dc816ad3821033.mspx)
[10] 'File in Use' Demystified (http://shellrevealed.com/blogs/shellblog/archive/2007/03/29/Your-File-Is-In-Use_2620_-Demystified.aspx)
[11] MSDN: Property System (http://msdn2.microsoft.com/en-us/library/aa969361.aspx)
[12] IColumnProvider Interface on MSDN (http://msdn2.microsoft.com/en-us/library/bb776147.aspx)
[14] Creating Context Menu Handlers (http://msdn.microsoft.com/en-us/library/cc144171(VS.85).aspx)
[17] Thumbnail Handlers (http://msdn.microsoft.com/en-us/library/cc144118(VS.85).aspx)
[18] .NET Shell extension from Microsoft – Photo Info tool (http://www.microsoft.com/downloads/details.aspx?familyid=B038D4B5-1D88-437C-9F54-1FB0D210B5EF&displaylang=en)
[19] MSDN Mag (http://msdn.microsoft.com/msdnmag/issues/04/01/WindowsShell/default.aspx#edupdate)
[20] Don't do Shell Extension Handlers in .NET (http://blogs.msdn.com/junfeng/archive/2005/11/18/494572.aspx)

External links

- Sullivan, Kent. "The Windows 95 User Interface: A Case Study in Usability Engineering" (http://www.sigchi.org/chi96/proceedings/desbrief/Sullivan/kds_txt.htm). © 1996 for Association for Computing Machinery, Inc. (http://www.acm.org) An article written by Kent Sullivan, a developer on the Windows 95 UI team, detailing the usability studies and development processes that led to the creation of the Windows 95 Explorer interface.
- How To Customize the Windows Explorer Views in Windows XP (http://support.microsoft.com/kb/307856/en)
- MSDN: Creating Shell Extension Handlers (http://msdn.microsoft.com/en-us/library/cc144067.aspx)
- The Complete Idiot's Guide to Writing Shell Extensions (http://www.codeproject.com/shell/shellextguideindex.asp)
- Namespace extensions – the undocumented Windows Shell (http://www.codeproject.com/KB/shell/shlext.aspx)
- InfoTip shell extension in Delphi (http://delphi.about.com/library/bluc/text/uc071701a.htm)

Filesystem_permissions

Most current file systems have methods of assigning **permissions** or **access rights** to specific users and groups of users. These systems control the ability of the users to view or make changes to the contents of the filesystem.

Operating system variations

Unix-like and otherwise POSIX-compliant systems, including Linux-based systems and all Mac OS X versions, have a simple system for managing individual file permissions, which in this article are called "traditional Unix permissions". Most of these systems also support some kind of access control lists, either proprietary (old HP-UX ACLs, for example), or POSIX.1e ACLs, based on an early POSIX draft that was abandoned, or NFSv4 ACLs, which are part of the NFSv4 standard.

Microsoft and IBM DOS variants including MS-DOS, PC DOS, Windows 95, Windows 98, Windows 98 SE, and Windows Me) do not have permissions, only file attributes. There is a read-only attribute (R), which can be set or unset on a file by any user or program, and therefore does not prevent him/her from changing/deleting the file. There is no permission in these systems which would prevent a user from reading a file.

Other MS-DOS/PC DOS-compatible operating systems such as DR DOS 3.31 and higher, PalmDOS, Novell DOS, and OpenDOS, FlexOS, 4680 OS, 4690 OS, Concurrent DOS, Multiuser DOS, Datapac System Manager and IMS REAL/32 support read/write/execute/delete file/directory access permissions on FAT volumes. With the exception of FlexOS, 4680 OS, 4690 OS all these operating systems also support individual file/directory passwords. All operating systems except for DR DOS, PalmDOS, Novell DOS and OpenDOS also support three independent file/directory ownership classes world/group/owner, whereas the single-user operating systems DR DOS 6.0 and higher, PalmDOS, Novell DOS and OpenDOS only support them with an optional multi-user security module loaded.

OpenVMS (a.k.a. VMS), as well as Microsoft Windows NT and its derivatives (including Windows 2000 and Windows XP), use access control lists (ACLs)[1] to administer a more complex and varied set of permissions. OpenVMS also uses a permission scheme similar to that of Unix, but more complex. There are four categories (System, Owner, Group, and World) and four types of access permissions (Read, Write, Execute, and Delete). The categories are not mutually disjoint: World includes Group which in turn includes Owner. The System category independently includes system users (similar to superusers in Unix).[2]

Classic Mac Operating Systems are similar to DOS variants and DOS-based Windows: they do not support permissions, but only a "Protected" file attribute.

The AmigaOS Filesystem, AmigaDOS supports a relatively advanced permissions system, for a single-user OS. In AmigaOS 1.x, files had Archive, Read, Write, Execute and Delete (collectively known as ARWED) permissions/flags. In AmigaOS 2.x and higher, additional Hold, Script, and Pure permissions/flags were added.

Mac OS X versions 10.3 ("Panther") and prior use POSIX-compliant permissions. Mac OS X, beginning with version 10.4 ("Tiger"), also support the use of NFSv4 ACLs. They still support "traditional Unix permissions" as used in previous versions of Mac OS X, and the *Apple Mac OS X Server version 10.4+ File Services Administration Manual* recommends using only traditional Unix permissions if possible. It also still supports the Mac OS Classic's "Protected" attribute.

Solaris ACL support depends on the filesystem being used; older UFS filesystem supports POSIX.1e ACLs, while ZFS supports only NFSv4 ACLs.[3]

Linux supports POSIX.1e ACLs. There is experimental support for NFSv4 ACLs for ext3 filesystem.[4]

FreeBSD supports POSIX.1e ACLs on UFS, and NFSv4 ACLs on UFS and ZFS.[5]

IBM z/OS implements file security via RACF (Resource Access Control Facility)[6]

Traditional Unix permissions

Permissions on Unix-like systems are managed in three distinct *classes*. These classes are known as *user*, *group*, and *others*.

When a file is created on a Unix-like system, its permissions are restricted by the umask of the process that created it.

Classes

Files and directories are owned by a user. The owner determines the file's *owner class*. Distinct permissions apply to the owner.

Files and directories are assigned a group, which define the file's *group class*. Distinct permissions apply to members of the file's group members. The owner may be a member of the file's group.

Users who are not the owner, nor a member of the group, comprise a file's *others class*. Distinct permissions apply to others.

The *effective permissions* are determined based on the user's class. For example, the user who is the owner of the file will have the permissions given to the owner class regardless of the permissions assigned to the group class or others class.

Permissions

There are three specific permissions on Unix-like systems that apply to each class:

- The *read* permission grants the ability to read a file. When set for a directory, this permission grants the ability to read the **names** of files in the directory (but **not** to find out any further information about them such as contents, file type, size, ownership, permissions, etc.)
- The *write* permission grants the ability to modify a file. When set for a directory, this permission grants the ability to modify entries in the directory. This includes creating files, deleting files, and renaming files.
- The *execute* permission grants the ability to execute a file. This permission must be set for executable binaries (for example, a compiled C++ program) or shell scripts (for example, a Perl program) in order to allow the operating system to run them. When set for a directory, this permission grants the ability to access file contents and metainfo if its name is known, but not list files inside the directory (unless *read* is set).

The effect of setting the permissions on a directory (rather than a file) is "one of the most frequently misunderstood file permission issues".[7]

When a permission is not set, the rights it would grant are denied. Unlike ACL-based systems, permissions on a Unix-like system are not *inherited*. Files created within a directory will not necessarily have the same permissions as that directory.

Changing permission behavior with setuid, setgid, and sticky bits

Unix-like systems typically employ three additional modes. These are actually attributes but are referred to as permissions or modes. These special modes are for a file or directory overall, not by a class.

- The *set user ID*, *setuid*, or SUID mode. When a file with setuid is executed, the resulting process will assume the effective user ID given to the owner class. This enables users to be treated temporarily as root (or another user).
- The *set group ID*, *setgid*, or SGID permission. When a file with setgid is executed, the resulting process will assume the group ID given to the group class. When setgid is applied to a directory, new files and directories created under that directory will inherit the group from that directory. (Default behaviour is to use the primary group of the effective user when setting the group of new files and directories.)

- The *sticky* mode. (Also known as the *Text* mode.) The typical behaviour of the sticky bit on executable files encourages the kernel to retain the resulting process image in memory beyond termination. On a directory, the sticky permission prevents users from renaming, moving or deleting contained files owned by users other than themselves, even if they have write permission to the directory. Only the directory owner and superuser are exempt from this.

These additional modes are also referred to as *setuid bit*, *setgid bit*, and *sticky bit*, due to the fact that they each occupy only one bit.

Notation of traditional Unix permissions

Symbolic notation

There are several ways by which Unix permissions are represented.

The most common form is **symbolic notation** as shown by `ls -l`.

Three permission triads	
first triad	what the owner can do
second triad	what the group members can do
third triad	what other users can do
Each triad	
first character	`r`: readable
second character	`w`: writable
third character	`x`: executable `s` or `t`: executable and setuid/setgid/sticky `S` or `T`: setuid/setgid or sticky, but not executable

The first character indicates the file type and is not related to permissions. The remaining nine characters are in three sets, each representing a class of permissions as three characters. The first set represents the *user* class. The second set represents the *group* class. The third set represents the *others* class.

Each of the three characters represent the read, write, and execute permissions:

- `r` if reading is permitted, `-` if it is not.
- `w` if writing is permitted, `-` if it is not.
- `x` if execution is permitted, `-` if it is not.

The following are some examples of symbolic notation:

- -rwxr-xr-x a regular file whose user class has full permissions and whose group and others classes have only the read and execute permissions.
- crw-rw-r-- a character special file whose user and group classes have the read and write permissions and whose others class has only the read permission.
- dr-x------ a directory whose user class has read and execute permissions and whose group and others classes have no permissions.

Note that in some Unix like systems:

- + (plus) suffix indicates an Access Control List which can grant additional permissions. See details with `man getfacl`.
- . (dot) suffix indicates an SELinux context is present. See details with ls -Z.

Numeric notation

Another method for representing Unix permissions is an octal (base-8) notation as shown by `stat -c %a`. This notation consists of at least three digits. Each of the three rightmost digits represents a different component of the permissions: owner, group, and others.

Each of these digits is the sum of its component bits (see also Binary numeral system). As a result, specific bits add to the sum as it is represented by a numeral:

- The read bit adds 4 to its total (in binary 100),
- The write bit adds 2 to its total (in binary 010), and
- The execute bit adds 1 to its total (in binary 001).

These values never produce ambiguous combinations; each sum represents a specific set of permissions. More technically, this is an octal representation of a bit field – each bit references a separate permission, and grouping 3 bits at a time in octal corresponds to grouping these permissions by user, group, and others.

These are the examples from the Symbolic notation section given in octal notation:

Symbolic Notation	Octal Notation	English
----------	0000	no permissions
---x--x--x	0111	execute
--w--w--w-	0222	write
--wx-wx-wx	0333	write & execute
-r--r--r--	0444	read
-r-xr-xr-x	0555	read & execute
-rw-rw-rw-	0666	read & write
-rwxrwxrwx	0777	read, write, & execute

User private group

Some systems diverge from the traditional POSIX-model of users and groups, by creating a new group – a "user private group" – for each user. The "user private group" scheme can be preferred for a variety of reasons[8] [9] [10] including using a umask of 002 and not having every "user" able to write to newly created files. In this case however, no other users must be added to the "user private group" or they will have write-permission on all files.

See also

- Access control list
- chattr
- chmod, the command used to set permissions on Unix-like systems
- File system
- Group identifier (Unix)
- lsattr
- POSIX
- umask
- User identifier (Unix)

References

[1] http://technet.microsoft.com/en-us/library/bb727008.aspx
[2] http://h71000.www7.hp.com/doc/731final/6489/6489pro_025.html#int_prot_code
[3] http://docs.sun.com/app/docs/doc/819-5461/ftyxi?a=view
[4] http://www.suse.de/~agruen/nfs4acl/
[5] http://wiki.freebsd.org/NFSv4_ACLs
[6] http://publib.boulder.ibm.com/infocenter/zos/v1r12/index.jsp?topic=%2Fcom.ibm.zos.r12.e0ze100%2Fracf.htm
[7] Hatch, Brian. "Linux File Permission Confusion pt 2" (http://www.hackinglinuxexposed.com/articles/20030424.html), "Hacking Linux Exposed", April 24, 2003, accessed July 6, 2011.
[8] O'Reilly ONLamp Blog, Using User Private Groups (http://www.oreillynet.com/onlamp/blog/2006/09/using_user_private_groups.html)
[9] Red Hat 9 Manual, User Private Groups (http://www.redhat.com/docs/manuals/linux/RHL-9-Manual/ref-guide/s1-users-groups-private-groups.html)
[10] Red Hat Enterprise Linux 5 Manual, 32.5. User Private Groups (http://www.redhat.com/docs/manuals/enterprise/RHEL-5-manual/Deployment_Guide-en-US/s1-users-groups-private-groups.html)

External links

- Apple Mac OS X Server version 10.4+ *File Services Administration* Manual (see pages 16-26) (http://manuals.info.apple.com/en/File_Services_v10.4.pdf)
- "Linux File Permission Confusion" (http://www.hackinglinuxexposed.com/articles/20030417.html) by Brian Hatch 2003.
- "Linux File Permission Confusion pt 2" (http://www.hackinglinuxexposed.com/articles/20030424.html) by Brian Hatch 2003.
- The Linux Cookbook: Groups and How to Work in Them (http://dsl.org/cookbook/cookbook_9.html) by Michael Stutz 2004

File_server

In computing, a **file server** is a computer attached to a network that has the primary purpose of providing a location for shared disk access, i.e. shared storage of computer files (such as documents, sound files, photographs, movies, images, databases, etc.) that can be accessed by the workstations that are attached to the same computer network. The term *server* highlights the role of the machine in the client–server scheme, where the *clients* are the workstations using the storage. A file server is not intended to perform computational tasks, and does not run programs on behalf of its clients. It is designed primarily to enable the storage and retrieval of data while the computation is carried out by the workstations.

File servers are commonly found in schools and offices, where users use a LAN to connect their client computers.

Types of file servers

A file server may be dedicated or non-dedicated. A dedicated server is designed specifically for use as a file server, with workstations attached for reading and writing files and databases.

File servers may also be categorized by the method of access: Internet file servers are frequently accessed by File Transfer Protocol (FTP) or by HTTP (but are different from web servers, that often provide dynamic web content in addition to static files). Servers on a LAN are usually accessed by SMB/CIFS protocol (Windows and Unix-like) or NFS protocol (Unix-like systems).

Database servers, that provide access to a shared database via a database device driver, are *not* regarded as file servers.

Design of file servers

In modern businesses the design of file servers is complicated by competing demands for storage space, access speed, recoverability, ease of administration, security, and budget. This is further complicated by a constantly changing environment, where new hardware and technology rapidly obsolesces old equipment, and yet must seamlessly come online in a fashion compatible with the older machinery. To manage throughput, peak loads, and response time, vendors may utilize queuing theory[1] to model how the combination of hardware and software will respond over various levels of demand. Servers may also employ dynamic load balancing scheme to distribute requests across various pieces of hardware.

The primary piece of hardware equipment for servers over the last couple of decades has proven to be the hard disk drive. Although other forms of storage are viable (such as magnetic tape and solid-state drives) disk drives have continued to offer the best fit for cost, performance, and capacity.

Storage

Since the crucial function of a file server is storage, technology has been developed to operate multiple disk drives together as a team, forming a disk array. A disk array typically has cache (temporary memory storage that is faster than the magnetic disks), as well as advanced functions like RAID and storage virtualization. Typically disk arrays increase level of availability by using redundant components other than RAID, such as power supplies. Disk arrays may be consolidated or virtualized in a SAN.

Network-attached storage

Network-attached storage (NAS) is file-level computer data storage connected to a computer network providing data access to heterogeneous clients. NAS devices specifically are distinguished from file servers generally in a NAS being a computer appliance – a specialized computer built from the ground up for serving files – rather than a general purpose computer being used for serving files (possibly with other functions). In discussions of NASs, the term "file server" generally stands for a contrasting term, referring to general purpose computers only.

As of 2010 NAS devices are gaining popularity, offering a convenient method for sharing files between multiple computers.[2] Potential benefits of network-attached storage, compared to non-dedicated file servers, include faster data access, easier administration, and simple configuration.[3]

NAS systems are networked appliances containing one or more hard drives, often arranged into logical, redundant storage containers or RAID arrays. Network Attached Storage removes the responsibility of file serving from other servers on the network. They typically provide access to files using network file sharing protocols such as NFS, SMB/CIFS (Server Message Block/Common Internet File System), or AFP.

Security

File servers generally offer some form of system security to limit access to files to specific users or groups. In large organizations, this is a task usually delegated to what is known as directory services such as openLDAP, Novell's eDirectory or Microsoft's Active Directory.

These servers work within the hierarchical computing environment which treat users, computers, applications and files as distinct but related entities on the network and grant access based on user or group credentials. In many cases, the directory service spans many file servers, potentially hundreds for large organizations. In the past, and in smaller organizations, authentication can take place directly to the server itself.

See also

- Backup
- File Transfer Protocol (FTP)
- Server Message Block (SMB)
- Network-attached storage (NAS)
- Enterprise content management

References

[1] File and Work Transfers in Cyclic Queue Systems, D. Sarkar and W. I. Zangwill, Management Science, Vol. 38, No. 10 (Oct., 1992), pp. 1510-1523

[2] CDRLab Test (http://cdrlab.pl/article_strona_9571_1.html) (in Polish)

[3] InfoStor. NAS Advantages: A VARs View (http://www.infostor.com/index/articles/display/55961/articles/infostor/volume-2/issue-4/news-analysis-trends/nas-advantages-a-vars-view.html), April 01, 1998. By Ron Levine.

Article Sources and Contributors

File_Manager_(Windows) *Source*: http://en.wikipedia.org/w/index.php?title=File_Manager_%28Windows%29 *Contributors*: A876, AlistairMcMillan, Andareed, Bhowden, Bubble94, Crystallina, Damian Yerrick, Dojarca, Gobbleswoggler, GrandPoohBah, Guy Harris, Harryboyles, Heqs, Josh the Nerd, Kukini, Linuxgal, Lofote, MDGx, Pol098, Quietbritishjim, RaviC, RingtailedFox, Rjwilmsi, Rmsuperstar99, Ryan Norton, SF007, Sfan00 IMG, Sikon, SpLoT, Srice13, Trusilver, Uncle Milty, WOSlinker, Warren, Welsh, Wendy.krieger, WikHead, 33 anonymous edits

File_manager *Source*: http://en.wikipedia.org/w/index.php?title=File_manager *Contributors*: 16@r, 5 albert square, 65.3.231.xxx, AlexSedov, AlistairMcMillan, Alvin Seville, Amethyst1234, Angrytoast, Antandrus, Apachegila, Archer1742, Arepo, Autoerrant, Beland, Big Bob the Finder, Borgx, Braincricket, Bratch, Charitwo, Chiprunner, ChrisTracy, Coldacid, Conversion script, Creidieki, Damian Yerrick, David Rolek, Dbenzhuser, Dcljr, Dekisugi, Djegan, Dlpkbr, Doc aberdeen, Dojarca, Dolukhanov, Don Cuan, Don4of4, Dr Roots, DrPaul0401, EdC, Elsendero, Emperorbma, Eresus, Exidor, F.j.sant, Fanaticdestroyer, Feydey, Flapdragon, Fleepkay, FleetCommand, Frap, Gamer9832, GeorgeHara, Gerbolya, Ghettoblaster, Guy Harris, Guy Peters, Hapsiainen, Havarn, Helix84, Heqs, Hhielscher, Hoary, Homer Landskirty, Honette, Honta, Hulten, Iheartlondon, JamesBWatson, Janto, Jarry1250, Jengelh, Jerryobject, Jman, Jno, Johnbod, Josh the Nerd, KAtremer, Karnesky, Karrmann, Kayamon, Kiand, Kievite, Kimero, Kinu, Little Professor, Loginnigol, LordRM, Lotje, Lpenz, M Ahmad Noori, MCBastos, Mark Arsten, MegamanX64, Melrin, Michaeloqu, Minghong, Mmj, Monedula, Morph78, Muffinmanpoo, Mver, Mwtoews, NJJ.Rocher, Nakon, NantucketNoon, Nesmojtarkala, Norm, OlEnglish, OlavN, Oneiros, Ozysoft, P0lyglut, Paulmnguyen, Pdfpdf, Peterkyc, Pmsyyz, Pomte, Putte, PutzfetzenORG, Qxz, RScheiber, RaccoonFox, Rettetast, Rich Janis, Robbie098, Rpm13, Sanjeevverma2011, Semsi Paco Virchow, Sheps11, Sin-man, Siracusa, Smart170, Snek01, Sphaira, SteveSims, Suffusion of Yellow, Sun Creator, Survivor2Success!, Sysnav, Tds247, Tedickey, Temarez, TenPoundHammer, TheParanoidOne, Thv, Tobias Bergemann, Trivialist, Ugen64, Vor0nwe, Wackymacs, Whitebox, WinNc Developer, Windharp, Wknight94, Xmoogle, Yardie Lobo, Yug, 245 anonymous edits

Microsoft_Windows *Source*: http://en.wikipedia.org/w/index.php?title=Microsoft_Windows *Contributors*: 03jmgibbens, 1(), 16@r, 1nt2, 2mcm, 2toy mora, 62.253.64.xxx, 67773732TYU, 9ms, A Raider Like Indiana, A gnome, A-giau, A7x, ACCOM2222, Aaleksanyants, Aanhorn, Abhishikt, Abletonlive05, AbsoluteFlatness, Academic Challenger, Acroterion, Adam Mirowski, Adamacious, Adamodell, Admin@pcrevs.com, Ae-a, Aeæ, Afro Article, Ageha Winds, Ahodacsek, Ahoerstemeier, Aidan W, Aido2002, Aihtdikh, Airplaneman, Ajm81, Akamad, Akhristov, Akira-otomo, Aksi great, Albert0057, Aldie, Alegoo92, Alemily, Alfio, AlistairMcMillan, Allstarecho, Almafeta, Alpha 4615, Alphathon, Alphius, Alshaheen15, Althepal, Am088, Amakuru, Amcfreely, Amrykid, Andre Engels, Andrevan, Andrewtechhelp, Andros 1337, Andy16666, Andyh2, Angeljon121, Ann Stouter, Anog, Anonymous Dissident, Anonymous56789, Antandrus, Anthall1991, Antimatter15, Antique Rose, Anville, Anþony, Aomarks, Aqair, Aranel, Arch dude, Archer3, Archivist, Aria1561, Arnoldkul, Arthena, Ashdurbat, AstroNomer, Astroview120mm, Atifniyaz, Aude, Audrius u, AussieLegend, AvantgardeMVC, Aveilleux, Avenue, Avihut, Avocado27, AwesomeSaucer9, AxG, Ayjay1545, Azrael Nightwalker, B, BNSF Man, BUzTeD, Badwolf2212, Bakery2k, Banes, Barek, Barrera marquez, Bartosz, Bbq man, Bbriggs1, Bdoserror, Bdshort1, Beao, Beinsane, Ben-Zin, BenBen1234, Benandorsqueaks, Benc, Berek, Bergsten, Betelgeuse, Bevo, Bhadani, Bibliomaniac15, Big Bird, Big Booger, BigCow, Biker Biker, BillG, BillWSmithJr, Billyswong, Bimmerosx, Binsurf, Binzisimpsons, Bissinger, Blackanddarkness, Blackcap95, Blackcats, Blakkandekka, Blaxthos, Bleakcomb, Blobglob, Blowdart, Blubberboy92, BlueCaper, Blueforce4116, Bluestriker, Bo98, Bobo192, Bobwrit, Bogods, Bollyjeff, Boogie314, Boris Allen, BostonMA, Boylett, Brandizzi, Brandon Brown, Brian0918, BrianGo28, BrianRecchia, Brianski, Brisvegas, BrokenSegue, Bubba hotep, Buchanan-Hermit, Buddha24, Bunnyhop11, Burntsauce, CRFWNY, CWenger, Cacophony, Cadiomals, Caffelice, Caltas, Camembert, Can't sleep, clown will eat me, Canadian-Bacon, Candamir, Candorwien, CanisRufus, Caper13, CaptainVindaloo, Casper10, Casper2k3, Cayindra, Cbrown1023, Cburnett, Ceeon, Celestra, CesarB, Ceyockey, Cff12345, Cgnabod, Cgs, Charles Gaudette, Charles dye, Chazz, Cherkash, Chille, Chmpoure, Chocolatemilk94, Choppie3000, Chowbok, Chris 73, Chris Pickett, Chris the speller, Christian List, Chriswoz, Ciao 90, CitronManden, Cityscape4, Cjcamilla, Cjordan93, ClamDip, ClockworkTroll, Clyde1998, Cmckain14, Cncccer, Cncxbox, Codename Lisa, Codificate, Codyblevins, Coffee, Colejohnson66, Colin Hill, Cometstyles, Commander Keane, CommonsDelinker, Compfreak7, Computer boy2, Computerdan000, ConCompS, Coniosis, Conman23456, Conversion script, Coolcaesar, CoolingGibbon, Corporal clegg48, Cosmicdense, Courcelles, Cp111, Cpiral, Crazycomputers, Crazyromo, Crem23, Cremepuff222, Cristan, Crpietschmann, Cryptic, Cstanners, Ctbolt, Cuvtixo, Cvinoth, Cwolfsheep, Cyfraw, Cyktsui, Cynical, Cyrius, D, D thadd, DHN, DOHC Holiday, DaDrumBum, DalekClock, Damian Yerrick, Dan D. Ric, Daniel5127, DanielRigal, Danny Beaudoin, Daranz, DarkFalls, DarkHorizon, Darranc, Darremon, Darthnader37, Dasani, DataMatrix, Databases, Davelane, David Biddulph, David Gerard, David1409, Davidjk, Davidvkimball, Dcandeto, Dcolvin, Dcshank, Deanhowell123, Debackerl, Deeahbz, Deepen03, Dehumanizer, Delirium, Delldot, Deltabeignet, Demmy, Denniss, Derek Parnell, Deryck Chan, Dethomas, Deus2, Dieboybun, Diegogrez, DigbyDalton, Digita, DigitalLife, Dimeji.97, Dimre01, Dina, Dinhtuydzao, Dinjired, Dinnerface, DinosaursLoveExistence, Dipankan001, Disorganisation Man, Dj789, Djegan, Djhybrid117, Djmasala, Dmerrill, DmitryKo, Dog1818, Dojarca, Dolphinn, Downloaddude1258, Dp462090, Dragar Gt, DragonLord, DragonflySixtyseven, Dragonhelmuk, Dstln, Dtech, Dudesleeper, Dungodung, Dust Filter, Dustin gayler, Dwheeler, Dylan Flaherty, Dysprosia, Dzubint, EH74DK, Eagleal, Ed g2s, Edgar181, Edgarde, EdgeOfEpsilon, Edward, Eelamstylez77, Eggsacute, Egil, Eivindsol, El Dominio, Elassint, ElectroPro, Eliotwiki, ElliotAdderton, Ellmist, Elm-39, Eloquence, Elwood00, Emacsuser, Emmette Hernandez Coleman, Enco1984, Endofskull, Enigmaaaaa, Enno, Enochlau, Ente75, Eraserhead1, Ergosteur, Esanchez7587, Escape Orbit, Esebi95, Evice, Evil Monkey, Evil saltine, Evosoho, Ewlyahoocom, Exodite, F80, FF2010, Fabsss, Faisal.akeel, Faithlessthewonderboy, Falco McCloud, Fantasy, Faradayplank, FatalError, Fatjoe151, Favonian, FayssalF, Fbv65edel, Fdp, Felipe Aira, Fennec, Ferkelparade, Ffx, Fiskegalen92, Fitzebwoy, Fiver2552, Flamurai, Flanakin, FleetCommand, Flying Bishop, FlyingPenguins, Foo1995, Fowl2, FrancoGG, Frap, Fraslet, Freakofnurture, Frecklefoot, Fred Bradstadt, Frederik.Questier, FreeKresge, Freeeekyyy, Frenchman113, FreplySpang, Frmorrison, FrozenPurpleCube, Frozenport, Fry1989, Fuck You, Funandtrvl, Furrykef, Fuzheado, G&CP, GDonato, GNUtoo, GSK, Gabrielpokemon, Gadfium, Gagetony8, Gaius Cornelius, Galactor213, Galoubet, Galwhaa, Gamer9832, GamerXp, Gamerzworld, Ganeshotaku, Ganfon, Gang65, Gaodifan, Gardar Rurak, Gary King, Gary Kirk, Gavin.perch, Gazpacho, Ged UK, Geek45, Geoffspear, George Adam Horváth, George Parastatidis, Georgia guy, Geraki, Gerbrant, Ghettoblaster, Gilbertogm, Gilliam, Glen, Gnepets, Goatasaur, Gobonobo, Gogo Dodo, Goldom, Googler459, Googlesucks56789, Gosub, Graham87, GrandPoohBah, Grayshi, Grin, Grm wnr, Groggy Dice, Gronky, Grunt, Gryllida, Gscshoyru, Gtdp, Gunnar Guðvarðarson, Gustyfalcon, Guy Harris, Guyjohnston, HJ Mitchell, Haakon, Haeleth, Hahafatpeople, Hairchrm, Hall Monitor, Halsteadk, Hammersoft, Harryboyles, Haseo9999, Hbomb phd mom, Hcobb, Hdt83, Heapchk, Hebrides, Helixblue, Hello32020, Hendrixski, Henry W. Schmitt, HereToHelp, Heroman1234, Himanis Das, Hirosho, Holbred, Howardjp, Hungupbg, Hurtstotalktoyou, Husky, Hydr, Hypnoticcyst, I have wood, I8189720, IGod, ILOVELOL32, IMSoP, IXella007, Iain99, Ialsoagree, Ian Pitchford, Ianjones50, Ichangefacts, Ichthyoid, Iggymwangi, Ilaiho, Illegal Operation, Illyria05, Ilya, IlyaHaykinson, Ilyanep, Indefatigable, Indon, Injust, Interframe, Ioprwe, Ipad135, Ipodsocool, Iridescent, IsUsername, ItsProgrammable, Ivan Pozdeev, Ixfd64, J Di, J36miles, JCarriker, JIP, JLaTondre, JWSchmidt, JYOuyang, JYolkowski, Jacob Hnri 5, JacquesStrap, Jaericho, Jake Wasdin, Jamieostrich, Jan Hofmann, Jasper Deng, Jauerback, Jaxl, Jcbparry, Jcurtin, Jdlowery, Jdm64, Jeffwang, Jeh, JennKR, Jeremy Visser, JeremyA, Jerryseinfeld, Jerrysmp, Jesant13, Jesse V., Jesse Viviano, Jesus5555, Jesus764, Jesuss, JettaMann, Jevel66, Jiffy, Jigs41793, Jimmi Hugh, Jimothytrotter, Jimthing, Jiy, Jketola, Jkonline, Jmath666, Jmjglick, Jmoynihan08hm77, JoJaEpp, JoJan, JoThousand, Joanjoc, JoanneB, Joe2832, Joejoejo, Joffeloff, Johann Wolfgang, John, John Ericson, John Goettle, John Quincy Adding Machine, JohnJamesWIlson, JohnOwens, Johnclow13, Johnleemk, Johnmc, Johnuniq, Jojit fb, Jon vs, JonasL, Jonatan Swift, Jondel, Jonghyunchung, Jopxton, Jordan015, Jordanyoung17, Jose Concepcion, Josh Tumath, Josh the Nerd, Joshua Issac, JoshuaArgent, Joyous!, Jpoke89, Jrcure, Jsmethers, Jtc, Ju66l3r, JuJube, JuanC08, Julian Mendez, Jumbo Snails, Justanother, Justashuman960, Jyoz, Kakomu, Kar.ma, Karimarie, Karlo2002, Karnesky, Kaycubs, Kaysov, Kbh3rd, Kbolino, Kelly Martin, Kelton2, Kenny Strawn, Kesla, Kevin12xd, Kevinkor2, Keyser Söze, Khym Chanur, Kierenj, Kigali1, KingpinE7, Kinneyboy90, Kipholbeck, Kiralexis, Kissmeplease, Kittyhawk2, Kjhonsa, Kkm010, Klingoncowboy4, KnowledgeOfSelf, Knutux, Kohlmalo, Kolonuk, Koman90, Kop, Korossyl, Kozuch, KrakatoaKatie, Krawi, KrisBogdanov, Krystaleen, KumardipSarkar, Kungfuadam, Kungming2, Kurieeto, Kushared, Kuyi123w, Kwen, Kypr8, L Kensington, La goutte de pluie, Lacrimosus, Lahiru k, Lantay77, Larry laptop, Lbs6380, Lcarsdata, Lead$peaker, Leandrod, Lectonar, LeeHunter, Leif, Leithp, LeoNomis, Letdorf, Leuko, Liberty Miller, Lightmouse, LilHelpa, Linear88, Linuxbeak, Linuxerist, LionKimbro, LittleBenW, Liudant, Llalala, Localzuk, Loginnigol, Loismum, Lonaowna, LonelyBeacon, Looksliketrent, Lord British, Lordalpha1, Lotje, Lou.weird, Lowellian, Lprd2007, Luce007, Luk, Lumpbucket, Luna Santin, Lupin, M Johnson, MADCastro2012, MCC, MER-C, MERC, MJ94, MJGR, MK8, MMMEEE, MZMcBride, Ma8thew, Mac, MacGeekGuy, Macurry, Madcow 93, Maester mensch, Magister Mathematicae, Mailer diablo, MainFrame, Malo, Man123123, Marceki111, Marcok, Marcus Qwertyus, Mark, Mark85296341, MarkGallagher, MarkSpearmint, Markhurd, Markie2, Markustwofour, Martarius, Martyx, Marudubshinki, Marx Gomes, Marysunshine, Massysett, MasterCole, Mastershake86, Mathwizxp, Matt Crypto, Matt.forestpath, Matt99clancy, MattGiuca, Mattarata, Matthiaspaul, Mav, Mawcowboybillsbrick7, Max Schwarz, Maxim Masiutin, Maximus Rex, McLovin34, Mcwatson, Meelar, Megamanfan3, Mehran, Metalim, Methanegas, Metz2000, Mguy77, Mgw854, Michael Angelkovich, Michael Frind, Michael Ray, Michaeldadmum, Michaelpremsrirat, MichealH, Midgrid, Midkay, Midnight Green, Mike4ty4, MikeZuniga, Mild Bill Hiccup, Milfman, Minesweeper, MinorItem, Mirddes, Miros 0571, Mithent, Mjpieters, Mkbashier, Mkdw, Mlscdi, Mmoll21m, Mmoller, Mms, Mo0, Modster, Mokk123, Monedula, Monothleft, MontyB, Moonknightus, Moonridersaregay, Moonwolf14, Mr Bartels, Mr White, Mr. Lefty, Mr. XYZ, MrTechNews123, Mrdelayer, Mrleewilliams, Ms2ger, Mugunth Kumar, Mulad, Mushroom, Mvmarier, Mwtoews, Mxn, Myanw, NE Ent, NGSF, Nachmore, Naddy, Naltrexone, Nanshu, NapoliRoma, Nathan nfm, Nathanlilienthal13, Ne0Freedom, Neckername, Neelchauhan, Neo The User, Neo2256, NewGuy4, Newfraferz87, Nhellviktor, Nick2253, Nickenzi, Nicolas Love, Nimc, Nintendog master 54, Nishkid64, Nixdorf, Njan, Nmnmnm, NotinREALITY, Noypi380, Nozomimous, Nukleartoaster, Nurg, O, Odd bloke, Off!, Okwestern, OlEnglish, Olioster, Ollie the Magic Skater, Omicronpersei8, Onco p53, Ondrejsv, Oneupthextraman, Opagecrtr, Opelio, Optakeover, OregonD00d, Oscarthecat, Ost316, OtherPerson, Otisfrog, Otter Mii-kun, OwenX, Oxymoron83, Ozherb, PPGMD, Packard Bell Legend, Padsquad43, Pai Walisongo, Pakaran, Patrick, Patrickweeks007, Patrolman89, Paul Stansifer, Paulfp, Pavel Vozenilek, Pcuser42, Persian Poet Gal, Person1936, Petaluma Paranormal, Peter Grey, Peteturtle, Petrwiki, Peyman4u, Peyre, Pgiii, Pgk, Phanton, PhelperRF, Phil Sandifer, Philip Trueman, Phoenixlogan123, Phoogenb, Pigman, Pigsonthewing, Pikachu Bros., Pile0nades, Pinikas, Piratesmack, Plugwash, PoWwOw123, Pokyrek, Poor Yorick, Poweroid, Prjrv787, Prolog, Propound, Pschulz01, PsyMar, Puffin, Puramyun31, Pursey, QUAZWRATH, Qudder, Quinsareth, Quoladdie, Qwerty124gg, Qwitchibo, Qx132, Qxz, R'n'B, RB972, RTC, RadioActive, RadioFan, Radon210, Raja99, Ramanpotential, RandomP, Rasmasyean, Rbuj, Rchamberlain, Rcmouse1010, Rd syringe, Rderijcke, Rdsmith4, Reallikeunreal, Rebroad, Red Dalek, Red Director, RedKlonoa, RedXII, Redvers, Reilly, Reisio, Remember the dot, Repetition, Resplendent, RevolverOcelotX, Rezonansowy, Rgbc2000, Rhesusmonkeyboy, Rhobite, Rhtc, Rich Farmbrough, Richard Lotspard, Richard626, Ricjl, RickK, Riluve, Rincewind32, Rjecina, Rjwilmsi, Rl, Rlevse, Robartin, Robbie940, Robert Buzink, Robert H, Robert Xia, RobertG, Robertd, Robinhw, Rocastelo, Rocket71048576, Rockymountains, Rodeo90, Roketjack, RolandH, Rolandhelper, Rollins83, Ronark, Rsantmann, Rsm99833, Running, Rursus, Rutherfordjigsaw, Ruud Koot, Rwwww, Ryan Norton, Ryan Postlethwaite, Ryan t moua, S0aasdf2sf, SF007, SHARD, SHeumann, SNIyer12, ST47, SYSS Mouse, Sae1962, Sam Hocevar, Samuel, Samuel Blanning, Samvscat, Sango123, Sasquatch, Sauronjim, Scepia, SchmuckyTheCat, SchuminWeb, Science4sail, Sciencegeek10188, Scientus, Scifiintel, Sciurinæ, Scorpiona, Scottymoze, Scriberius, Scuiqui fox, SeanMack, Sebmathews, Sebrat, Secretlondon, Selfdiscipline, Selivanow, September 11 terrorist, Sethoeph, Sfan00 IMG, Shadowfyren, Shadowjams, Shakumafu, Shandris, Shanes, Shashank Shekhar, Shelmac, Sherylanne, Shibboleth,

Shinjiman, Shirudo, Shnout, Shwangtianyuan, Sigma 7, Sigmundpetersen, Sikon, Silvergoat, SimonEast, SimonP, SimonTrew, Simple Bob, Simxp, SirGrant, Siroxo, Sjakkalle, Sjorford, Skirks, Skiwi, Skyeap, Slashuer, Slathering, Sleepeeg3, Slipknotmetal, SmackEater, Smokizzy, SmoothNikola, Smurfy, Smyth, Snaxe920, Snoyes, SoSaysChappy, Solphusion, Somaditya, Somebody in the WWW, Soumyasch, SouthernNights, Southpark20, Spe88, Spear of fire, SpecMode, Specter01010, SpeedyGonsales, Spencerperry, Sperling, Spliced, Spookfish, Spug, Sri555666, Srnelson, Ssd, Starionwolf, Steel, Steezly, Stephanwehner, StephenH, Stephenchou0722, Steve2011, Stevenh123, Stevenmitchell, Stevietheman, Stilroc, Stino v, Stonda, StuIsCool, StuThomas, SudoGhost, Sumesh Dugar, Supercooljs2, Supers, Supremeknowledge, Suruena, SusanLesch, Susvolans, Swabjob, Sweetback, Swollib, Swotboy2000, SyntaxError55, Syrthiss, Szhang21, TAG.Odessa, TJ Spyke, Ta bu shi da yu, Tabletop, Tacvek, Tad Lincoln, Tadas12, TakuyaMurata, Tangotango, Tannin, Tarashav, Tawker, Tbhotch, Teamcritical, Tech2blog, Techie007, Techman224, Telecineguy, Template namespace initialisation script, Tepidpond, Terence, Tfgbd, Thaek, Thalakan, Thavian, The Colclough, The Disco Times, The Epopt, The Fish, The Geneticist, The Rambling Man, The wub, TheAirplaneGuy, TheChampionMan1234, TheChrisD, TheDoober, TheGeneralUser, TheKMan, TheNewPhobia, ThePointblank, Thealexweb, Theazman1, Thegreenj, Themodernizer, Thetehror, Thewallowmaker, Thewikipedian, Thingg, Think outside the box, This, that and the other, Thisisntfake1, Thomas Larsen, Thompson.matthew, Thorpe, Throup, Thu, Thumperward, TiCPU, Tiggerjay, TimR, Timberframe, Timmeh, Timwi, Tinton5, ToasterOS, Toddcs, Tom Morris, Tomcage9, Tommy Irianto, Tompagenet, Tomy9510, Tonsofpcs, Tony1, Tothwolf, Towel401, Towsonu2003, Toyotaboy95, Tracer9999, Train2104, Treekids, Tregoweth, TrekMaster, Trevor MacInnis, Tripacer99, Trivialist, Trusilver, Tsunaminoai, Ttwaring, TubularWorld, TuneyLoon, Turionaltec, TuukkaH, Twernt, Twickline, TyA, Tyomitch, Tyraios, Tyrol5, UMC2, Uli, Ultimus, Umofomia, UncleBubba, Unimaginative Username, Unixman83, Unknownperson1234, Uriel8, Utcursch, Vahid83, Valenciano, Vanished user 5zariu3jisj0j4irj, Vbrtrnn, Vedek Dukat, Vegaswikian, Vincent.premysler, ViperSnake151, Vontafeijos, Vrenator, W1 m2, Wackymacs, Walter Görlitz, WalterGR, Wanderson9, Wangmike, Wapcaplet, Warlordwolf, Warren, Weararedjacket, Wengier, Werdan7, Werideatdusk33, Westley Turner, Whiteford8, Whkoh, Widefox, Wiki alf, Wiki fanatic, WikiBone, WikiFew, WikiMan44, WikiRedactor, Wikidexel2, Wikinger, WikipedianMarlith, Wikiwaka101, Windowsknowitall@msn.com, WindowzRULZlolZ, Windsok, Wknight94, Wlklpedla is meant to be vandalized, Wmahan, Wmplayer, Woohookitty, Wow1000, Wragge, Wsoder, Wykis, X1987x, XJaM, Xation, Xelgen, Xeysz, Xgmx, Ximian99, Xlation, Xmachina, Xp54321, Xpclient, XrXeJoeXaXpXeXr, Xtreme racer, Xxglennxx, YUL89YYZ, Yama, Yamamoto Ichiro, Yamla, Yasirniazkhan, Yeeshenhao, Yehiahassan, Yergizmo, Ynhockey, Yousifnet, Yuckfoo, Yuhong, Z.E.R.O., Z220info, Z98, ZFU738, Zakfleming, Zapvet, Zarex, Zeeboid, Zepheus, Zer0431, Zntrip, Zondor, Zoney, Zundark, , , , 1542 anonymous edits

MS-DOS *Source*: http://en.wikipedia.org/w/index.php?title=MS-DOS *Contributors*: 15turnsm, 208.164.148.xxx, 2600:1014:B12E:937:0:0:0:103, 5 albert square, @modi, A Raider Like Indiana, ABCD, AVRS, Aacini, Aanhorn, Acdx, Achowat, Ae-a, After Midnight, Ahmad87, Ahodes1, Akhristov, Alansohn, Alanthehat, Aldie, Aleenf1, AlefZet, Alerante, Alex230592, Alexander Mclean, Alexwcovington, AlistairMcMillan, Allen4names, Altrio, An-chan, Anby, Andre Engels, Andrevan, Andrew Jackson, Andrew Stiff, Andrewpmk, Andy16666, AngelOfSadness, Animesh.saxena, Annavoskanyan, AnonMoos, Anss123, Anubaph, Anville, Arbor to SJ, Arborinus Verginix, Ardonik, Asdfasdfg23, Ashley Pomeroy, Asmpgmr, AussieLegend, Avouac, Avs5221, Ayecee, B4light, Babedacus, Ballon845, Beinsane, Ben-Zin, Betax, Bevo, Bfissa, Bilby, Billposer, Binsurf, Bkell, Blackcats, Blainster, Blowdart, Bobblewik, Bobo192, Boccobrock, Boxstaa, BradBeattie, Brooktree, Buddha24, CCFreak2K, CanisRufus, Capricorn42, CatherineMunro, Charl0929, Chemturion, Chitrapa, ChrisGualtieri, Clarkbhm, Cliftonmullins, Cobaltbluetony, Codename Lisa, CombatCraig, Conversion script, CoolD, CoolFox, Cooljrt, Crispmuncher, Critic-at-Arms, CudllllyCakeMixCore, Cwolfsheep, Czeror, DARTH SIDIOUS 2, DOSGuy, DSRH, DVD R W, Damian Yerrick, Dan100, Dave Farquhar, Dawnseeker2000, Deathawk, Deflective, Defydaydryers, Deisenbe, Demi, DerHexer, Dgies, Dgiraffes, Dirac1933, Dirk P Broer, Diverman, Djegan, Dkid3404, Dmsar, Doc glasgow, Dogman15, Dom Lochet, Donreed, Drat, Dudley Doright, Dureo, Dwilso, Dzubint, EALacey, Echo95, Edmundwoods, El Raki, Emperorbma, Emurphy42, Enric Naval, Epbr123, Epson291, Ericd, EthanL, Evan Russell, Evice, Evil saltine, Excirial, Fabartus, Family Guy Guy, Faradayplank, Favonian, Filzstift, Fonzzee, Foxroz, Freddy Riegel, Freedom to share, FreelanceWizard, Funkmasterj1, Funllama680, Furrykef, GL1zdA, GSMR, GVnayR, Gadfium, Gail, Galoubet, Gamingnews, Garik, Gasgiant69, Gazpacho, Gblaz, Geneb1955, Geoffrey, GermanX, Ghettoblaster, Gilliam, Gniw, Gogo Dodo, Goplat, Gotgod86, Gracenotes, GraemeL, Graham87, GrahamColm, Graue, Greenrd, Guliolopez, Gurch, Guy Harris, Hadal, Hairy Dude, Hannes Hirzel, Hayabusa future, Helixweb, Helpsloose, Herostratus, Hikmet483, Hyad, Hydrargyrum, Hypertime179, ITexter, Ian Fieggen, Illyria05, Infogaufire, Insanity Incarnate, Interiot, Interlude65, Intgr, Isaac Rabinovitch, Ixfd64, JSpung, Jaan513, Jagged 85, Jak123, JamesBWatson, Jamst98, Janizary, JarlaxleArtemis, Jascii, Jcam, Jeeves, JenKilmer, JeremyMcCracken, Jerryobject, Jffner, Jh51681, Jiddisch, Jmajeremy, Joffeloff, JohnnyMrNinja, Jojit fb, Josh the Nerd, Joshb306, Jrockley, JuJube, Julthep, Justayankeefan, Jwinters, Jzylstra, Kaiba, Karl-Henner, Kashmiri, Katalaveno, KelisFan2K5, Kephir, Kerjuk, Kesla, Kev19, Kevin B12, Khazar2, Kmweber, KnowledgeOfSelf, Koffieyahoo, Koman90, Konman72, Koopa turtle, Kozuch, Kwamikagami, Kwertii, Lakefall, Leandrod, Leojei, Lfwlfw, LilHelpa, Linkspamremover, Localzuk, Loismum, Lotje, Lproven, Lst27, Luigiacruz, MARQUIS111, MPS, Mabdul, Magus732, Mahjongg, Maralia, Marcodeo, Marek69, Martarius, Master Thief Garrett, MattGiuca, Mattgirling, Matthiaspaul, Mattisse, Maximus Rex, Mbssbs, Meemsinator, Meldor, Menneisyys, Methodius12, Michael Zimmermann, Michaelpremsrirat, Mike V, MikeRS, Mikeblas, Minesweeper, Misternuvistor, Modster, Molotovnight, Mr Pie Guy1234, Mr Tan, MrSomeone, Mrstonky, Mshonle, Mynameischarlesbarkley, Mysdaao, Mysid, N90p, NawlinWiki, Nbarth, Nepenthes, Neverstable, Nick Huitema, Nick Number, NickCT, Nintendough, Nk, Nono64, Norm mit, NoychoH, Nskbalu, Nv8200p, O.Koslowski, Ohms law, Oisjfw, Oliverlyc, Omicronpersei8, Oneupthextraman, OtherPerson, Oxymoron83, Ozzmosis, PCHS-NJROTC, PURAMILK, Pakaran, Palica, Patrick1982, Perfecto, Petersh, Petervee, Petzi1969, Peyre, PhilKnight, Pinikas, PlayStation 69, Pmurph5, Pointillist, Pol098, Posix memalign, PrintStar, Pykello, Qst, Quantumor, Qxz, R3m0t, RCX, Rack004, Rahulgr8888, Ranvel, Raven-14, Reaper Eternal, Rebroad, Redsoxcool, Redvers, Res2216firestar, Retron, RexNL, Ricer46, Rich Farmbrough, Riki, Roadstaa, RobLa, Robomaeyhem, RoyBoy, Rrburke, Ruud Koot, Rwwww, Ryanferber, Rynsaha, SF007, ST47, Saayiit, Saint-Paddy, Sam Korn, Samo00099, Sandstein, Scarian, SchmuckyTheCat, SchuminWeb, Schwarzes Nacht, Scootey, Scott Paeth, Seahorseruler, SeeSchloss, Shanedidona, Shell Kinney, Sigma 7, Sikon, SirClaymore, Skier Dude, Skizzik, Slakr, Slamminsam069, Smallman12q, Smyth, Sonamunda chaks, South Bay, SpLoT, Sri555777, Sshennar, Starionwolf, Stephen Gilbert, Stephenb, Stephenkall, Steven jones, Stewartadcock, Stovetopcookies, Stuart P. Bentley, Stwalkerster, Supaman89, Ta bu shi da yu, TakuyaMurata, Tamiera, Tangotango, Tannin, Tarquin, Tedickey, Tempshill, The Utahraptor, TheStarman, Thecheesykid, Thehelpfulone, Think outside the box, ThomasHarte, Tim Starling, Timharwoodx, Tobias Hoevekamp, Tompagenet, Tothwolf, TrainTC, Trlkly, Troyoda1990, TruckMonkey, Tyler, Tyomitch, UncleBubba, Uriyan, Useight, Valhalla, Vanished user 9i39j3, Versageek, Viajero, Vina, Vincent Lextrait, Vlad, Vpovilaitis, W8TVI, Wackymacs, Waggers, Warpozio, Warren, Wavelength, Wayne Slam, Wayward, Wbm1058, Wendy.krieger, Wengier, Wernher, Wgw2024, WhisperToMe, Whkoh, Whomp, Wiki alf, Wikievil666, Wiktator, Will Beback Auto, WillOakland, William Avery, Wimt, WinUser, Woohookitty, Wtshymanski, X3J11, Xeos, YK Times, Yourmomblah, Yuhong, Yunaffx, Zebs, Zer0431, Zzuuzz, אמס״סה123, 794 anonymous edits

Windows_95 *Source*: http://en.wikipedia.org/w/index.php?title=Windows_95 *Contributors*: 2602:306:3159:1270:255D:8606:446C:496A, 4twenty42o, 62.2.17.xxx, A Raider Like Indiana, Aaleksanyants, AcidHelmNun, Adrianwo, Aegicen, Aeharding, Afiod, Airplaneman, Akhristov, Alansohn, Alanthehat, Alaudo, Alden Jones, AlistairMcMillan, Altenmann, Amicon, Andre Engels, Aranea Mortem, Astroview120mm, Aths, Atomician, AussieLegend, BWCNY, Beeawwb, Bevo, Bige1977, Biruitorul, BitterTwitter, Blainster, Blukens, Bobianite, Bobo The Ninja, Bobo192, Bongwarrior, Bonkers191, Borisborf, Brian Kendig, Brion VIBBER, Browserview, Brucethemoose, Bryan Derksen, Bsadowski1, Bubba73, Buchanan-Hermit, Burger32, C2r, C:Amie, Callanecc, Caltas, CanisRufus, Carmichael, CenSorShot, Chapiown, Charles dye, CharlotteWebb, Chris16447, ChrisGualtieri, Chrisbolt, ChristTrekker, Chuunen Baka, Cj1340, Cobratank, Codename Lisa, Codyblevins, CommonsDelinker, Conversion script, Cosmogoblin, Cprompt, Csiew23, CyberSkull, Cyktsui, Dabomb87, DadaNeem, Daikiki, Dan6hell66, Dancraggs, Daniel Olsen, Daniel5127, DanielRigal, DarkShroom, Darklilac, Darrien, DaveMcLain, Davidbspalding, Dawnseeker2000, Dcirovic, Defixio, Dekisugi, Delirium, Denisutku, Denniss, DerHexer, Derekbd, Di Stroppo, Diannaa, Diego Moya, Digita, Discospinster, Diverman, Djegan, Dlohcierekim's sock, Doc Strange, Dojarca, Domesticenginerd, Dominar Rygel XVI, Dooties, Dorsal Axe, DragonHawk, Dugwiki, Dysprosia, Edex67, EditorE, Edokter, Edward, El C, Elmolovesyou1, Energybender, Ente75, EoGuy, Erc, Ericd, Eskovan, Euyyn, Evan Russell, Evice, Evil saltine, Excirial, Exozero, ExplicitImplicity, FF2010, FayssalF, Felipe Aira, Fernvale, FiP, Fishnet37222, Flameeyes, FleetCommand, FlyingPenguins, Foofy, Fourthords, Fribbler, Futuremonkey, Futurix, G&CP, GCarty, GL1zdA, GSK, Gaius Cornelius, Garethhamilton, Gaurav, General Heed, Gogo Dodo, GoldDragon, Grafen, Graham87, GrimRepr39, Grunt, Guanaco, Guy Harris, Hairy Dude, HappyDog, Happysailor, Hardylane, Harryboyles, Hasiko, Hayabusa future, Hdt83, Henrik, Henriok, Hidbaty223, Hippo X, Hohho56oy, Holothurion, Holpenthadoor, HumphreyW, Hwebers, I'm not crazy 'cause I take the right pills every day., I8189720, Ichthyoid, Inimbrium, Insanity Incarnate, Iridescent, Iuhkjhk87y678, Ivan Štambuk, Izcool, J Di, J.delanoy, JLaTondre, JRM, JamesBWatson, Jasonbluefire, Jasper Deng, Jc iindyysgvxc, Jdlowery, Jeffrey Mall, Jeffro77, Jeh, Jengelh, Jesperborgstrup, Jhdezjr, Jiang, Jim1138, Jjjnjjj, Jjproductionscorp, Jmarshmallow, Joanjoc, Joffeloff, Jon1412, Josh the Nerd, Jtalledo, Jwsolutions, Jéské Couriano, Ka Faraq Gatri, Kajervi, Katalaveno, Kate, Kesla, King Mustard, Kitch, Kiwikid32, Kjmitchell, Klow, Konman72, Kozuch, Krabben, Krazymike, Krich, Kubigula, Kurykh, Kutulu, Kylu, Kyng, LandSeaLock, Landroo, Letdorf, Liliana-60, Linuxerist, Linuxlove8088, Lloydsd, Localzuk, Lotje, Lugia2453, Luigiacruz, Luna Santin, Lupin, Lupine Proletariat, MER-C, Machina.sapiens, Mackatacka123, Madd the sane, Magus732, ManWithNoName, Manop, Martarius, Mathwiz593, MattGiuca, Matthiaspaul, Maxik, McSly, Mdwh, Megaman en m, Microwindowsxp1, Mike Rosoft, Mike65535, MikeRS, Mikko Paananen, Mild Bill Hiccup, MilesProwler, Minesweeper, Mogism, Mokwella, Moreau36, Mozzerati, Mr. Strong Bad, Mr.Z-man, MrJones, MrThomas369, Ms2ger, Mschlindwein, Msikma, Mushroom, MustardKing, Mxcatania, Mygerardromance, N5iln, Nanshu, Neelchauhan, NeilSantos, Nick Wilson, Nicolas Love, Nightwheel, NikiAnna, Nine Tail Fox, Nintendude, Nookls, Noozgroop, NotAnonymous0, Ofathens, Ohconfucius, Ohms law, Oknazevad, Oliverdl, Omicronpersei8, Ou tis, Oxymoron83, PainMan, Parkerbudzinski, Paul kenneth mora, PaulGS, Penhisa, Penmachine, Per Abrahamsen, Philipp Kern, Piano non troppo, Pilaf, Plasticbadge, Plugwash, Poccil, Poor Yorick, Posix memalign, ProhibitOnions, Psyk0, Public Menace, Punctured Bicycle, Qaywsxedc, Querswpoia, Quishyquishy, Qxz, R'n'B, RA0808, RAMChYLD, Radon210, RattleMan, Razoproductions, Real NC, Rebroad, Rechlin, Red1530, Reisio, Remstar, Resplendent, Retired username, Rgiordano, RickK, Ringostar4life, Rjwilmsi, Robert Merkel, Robert Moyse, Rockstone35, Ross Uber, Royalguard11, Rrburke, Rror, Russell1903, Ruud Koot, Rwwww, Ryan Norton, Ryūkotsusei, SF007, SHeumann, SJP, SNIyer12, ST47, Sakurambo, Samuel, Samuel Blanning, Samvscat, Satellizer, Scarian, SchmuckyTheCat, SchuminWeb, Scofield Boy, ScottSteiner, Seano1, Secant1, Sherkhon, Shjacks45, Sigma 7, Skizzik, Skybon, Sl, Smilesfozwood, Someguy1221, SpigotMap, Startstop123, Stefan2, Stephenchou0722, SteveSims, StuffOfInterest, Subwayguy, Swingballbros, Synchronism, Syrthiss, Szajd, T@Di, TRAiNER4, Ta bu shi da yu, Tannin, Tcncv, Template namespace initialisation script, Tgeairn, The Lovable Wolf, TheBertiebee, TheDataMonster, TheLoverofLove, TheREALCableGuy, TheRainstorn, Theaveng, Thingg, Timothyhouse1, Tjraptis, Tmopkisn, Tobi aa, Tommy2010, TonyW, Tralala23, Treisijs, TubularWorld, TuukkaH, Tyler McHenry, Tyomitch, Ulric1313, Unyoyega, Uriyan, Uusitunnus, Vague Rant, Vampire1995, Vegaswikian1, Vergency, Videoking, ViperSnake151, Vkem, Vmlinuz, WPSamson, Wackymacs, Waqas1987, Warren, Wdddqwedfqedf, Weedwhacker128, WeirdEars, WhisperToMe, Widefox, Wiki alf, WikiWriter67, WikipedianMarlith, Will Hawes, Willking1979, Windows55, Windowswindowswindowsxp, Wmplayer, Wolfcm, X686, Xpclient, Xslouis, Yakiv Gluck, Yidisheryid, Yintan, Yonatan2002, Yuhong, Yulu, Zabadab, Zidonuke, Zitbag 635, Zohebnensey, Δ, පසිඳු සාවිත්‍ර, , 762 anonymous edits

Windows_NT_4.0 *Source*: http://en.wikipedia.org/w/index.php?title=Windows_NT_4.0 *Contributors*: +u3)u!^ 7!3N, 4I7.4I7, A Raider Like Indiana, AS, Akhristov, Alansohn, Alexf, AlistairMcMillan, Already111, Angrytoast, Another-anomaly, AppleMacReporter, Araqsya Manukyan, Ashley Pomeroy, Atomice, AussieLegend, BS3k, BWCNY, Bebo2good1, Big Brother 1984, Bongwarrior, Bovineone, Bumm13, C:Amie, CharlotteWebb, Chowbok, Chris the speller, ChristTrekker, Chrysics, Codename Lisa, Colin Keigher, Copperoxide, Dan100, Deogene, Digita, DocWatson42, Dorsal Axe, Drawnmeets, Drmies, Dstruct2k, Edward321, EliasAlucard, Evil saltine, Father Goose, Felipe Aira, FleetCommand, Francisco Leandro, Frozenport, Furrykef, GCarty,

Gailwin, Gareth Griffith-Jones, Garyp01, Gikü, Gioto, Guy Harris, Haakon, Harryboyles, Hede2000, Hugo 87, Ibroadbent, Intellmac, Iuhkjhk87y678, IvanLanin, JamesBWatson, Jaruzel, Jasper Deng, Jc iindyysgvxc, Jdlowery, Jeh, Jeni, JetBlast, Jhdezjr, John of Reading, Jonathan Williams, Josephgkim, Josh the Nerd, Kauczuk, Kekship55, Koman90, Kozuch, Lasdlt, Letdorf, Lightmouse, Liliana-60, Luigiacruz, MDGx, MSTCrow, Mallardtheduck, Mardus, Marxo, Matthiaspaul, Mentifisto, Mike Rosoft, Minghong, Morte, NGSF, Nicolas Love, Noypi380, Ofus, Ozzmosis, PLCM, Paul kenneth mora, Paulfp, Peyre, Phils, RainbowOfLight, RattleMan, Rebroad, Rtyuire5, Rwwww, Ryankend, SF007, SHeumann, SchmuckyTheCat, Scholia, Scientus, Shlomital, Shuhao, Sigma 7, SimonEast, SpigotMap, Starkiller88, Ta bu shi da yu, Tangent747, Tgeairn, The Anonymouse, The High Fin Sperm Whale, Tohd8BohaithuGh1, TonyW, Tree Biting Conspiracy, Trusilver, Tyomitch, Typhoon, Vanished user ndufnwe98h23nc8hcskjf, Vegaswikian, Vegaswikian1, Videoking, ViperSnake151, WPSamson, Warren, Wereon, WikHead, WikipedianMarlith, Woohookitty, X686, Xpclient, Yshoulduknow, Yuhong, Zzyzx, , , 225 anonymous edits

Windows_Explorer *Source*: http://en.wikipedia.org/w/index.php?title=Windows_Explorer *Contributors*: -m-i-k-e-y-, AVM, AaThinker, Aaronw, Addshore, Akhristov, Alexei-ALXM, Algebra, AlistairMcMillan, Animum, Armageddon11, Astroview120mm, Atomice, AussieLegend, BMT, Bario1, Bart133, Bazzargh, Berntie, Berny68, Bhlj dfg vbfn kcbl gdvb`cnfmgbds bn, Blazaking, Bobo192, Bonadea, Bubbleboys, Buster2058, CaribDigita, Chowbok, ChrisJMoor, Clindhartsen, Codename Lisa, CommonsDelinker, Crystallina, Cuchullain, Cwolfsheep, CyberSach, CyberSkull, Czeror, DLoom, Dalibor Dragojevic, Daniel Simanek, DanielPharos, Davidbspalding, Dawright12, DeVerm, Digita, Dogmaticeclectic, Don Cuan, Dr.Kerr, Dustin gayler, EagleOne, Eric Kvaalen, EvilResident, Fastboy, FleetCommand, Fran z, Francois Trazzi, GSlicer, Gafaddict, Gagandeep, Gagetony8, Gamer9832, Gatul, Gazpacho, Ginkgo100, Ginsengbomb, Glenn W. Butler, Graham87, GreenReaper, Gunnar Hendrich, Harryboyles, Hayabusa future, Heron, Hideyuki, Himanis Das, Hux, Iggymwangi, Ilhanli, Imariano, Isaac Rabinovitch, Iste Praetor, Iteeluck, J.delanoy, JGXenite, Jasper Deng, Jeh, Jesant13, Jj0909jj, Josh Parris, Josh Tumath, José María Sola, Jschnur, Junkcops, KAMiKAZOW, Khazar, Kingal86, Kingpin13, Koman90, Krejcjus393, Kushal one, LeCire, Libcub, LinkTiger, LiquidFire, LittlegreenmanfromMars, Lkopeter, Loginnigol, Lotje, Mabdul, MacStep, Marcus Qwertyus, Matt714, Mediran, Menchi, Merlion444, Michaelas10, Mihai1025, Minesweeper, Minghong, Mipadi, NPrice, Nathan nfm, NazmusLabs, NeoChaosX, Nick2253, Noor1990, Norm, Patrick, Petri Krohn, Phatom87, Pol098, Pro Grape, Pronoun, Public Menace, Pxma, Radon210, Ralmin, Remember the dot, Requestion, Retodon8, Rezonansowy, Rtiae, Ryan Norton, SGBailey, SNIyer12, Scog, Senator2029, Sevvie, Shantavira, Sheogorath, Sietse Snel, Simcaster, Soumyasch, StalePhish, StefanB sv, Stephenchou0722, Stevage, Stib, Sun Creator, Szajd, Tagishsimon, Tamiera, The Thing That Should Not Be, TheBull1, TheFuzzball, Themodernizer, Theo10011, Timbernerslee, Tobias Bergemann, Toothpaste95, Trivelt, Trivialist, Ttony21, Tuanese, Tuntable, TutterMouse, Tyomitch, Ultimus, Vegaswikian, ViperSnake151, Waggers, Warpandas, Warren, Wimt, WulfTheSaxon, X25y en, X570, Xpclient, ZakuSage, ZephyrAnycon, Zundark, Володимир Ф, පසිඳු කාවින්ද, Ὁ οἶστρος, 243 anonymous edits

Filesystem_permissions *Source*: http://en.wikipedia.org/w/index.php?title=Filesystem_permissions *Contributors*: Addshore, AlistairMcMillan, Alvin-cs, Bachrach44, BenJWoodcroft, Black Kite, Bubba73, Calestyo, Casablanca2000in, Chenchaoyi, Chomwitt, DGerman, Daethnir, Daira Hopwood, Dcljr, Dekart, DreamGuy, FatalError, Fred Bradstadt, Furrykef, Gaius Cornelius, Galoubet, Gary King, Gbeeker, Gene91, Grawity, Greg McFarlane, HereToHelp, Hhielscher, Homberghp, Ian Kelling, Ignacioerrico, Intgr, IsUsername, Jahs, Jeliasen, Jengelh, Jerome Charles Potts, Joy, Jrtayloriv, Karl Dickman, Kbolino, Kpmlucid, LOL, Larry2rock, Lisamh, LockeShocke, Marko75, Matthiaspaul, Milan Keršláger, Mindmatrix, Moxfyre, Mqduck, Mr Gronk, Nbarth, NickPenguin, P0lyglut, PBS, Pandar, Perfection, Philip Trueman, Plugwash, Quaresti, Radagast83, Radiojon, Random832, Raysonho, Rob cowie, Rprpr, Rwxrwxrwx, Sam Hocevar, Samsara, Seidenstud, Sietse Snel, Simetrical, SimonP, Smyth, Snowolf, Spencerdcarlson, Spoon!, Squallwc, Swebert, Sweet xx, T-rex, Theshibboleth, Tobias Bergemann, Tony holden, Trasz, Unforgiven24, Unixguy, VMS Mosaic, Vadmium, Vitund, W1tgf, Wikipollock, Woody993, Xeleema, Xenoterracide, Zerohourrct, وخور, 113 anonymous edits

File_server *Source*: http://en.wikipedia.org/w/index.php?title=File_server *Contributors*: 16@r, 9258fahsflkh917fas, A8UDI, Adamsetzler, AhmedYounis, Altonbr, Arkuat, Bellemonde, Brandon5485, Butros, Calabe1992, CesarB, Christian List, Coredesat, DStoykov, Davidprior, Dr Gangrene, EdgeOfEpsilon, Emre D., Enchanter, Epbr123, Frap, Gardar Rurak, GermanX, Guillom, Hadrianheugh, Hairy Dude, Haseo9999, Husond, Intgr, Invenio, Issactang2, Japo, Jasper Deng, Jaxl, JeffC, Jklin, JoelDick, John, Joy, Kozuch, Kubanczyk, Leckley, Lightdarkness, LittleOldMe, Mange01, Marokwitz, Martpol, Maury Markowitz, Mets501, Moe Epsilon, Morning Sunshine, Mwtoews, Nbarth, Nicolas1981, Nquinnathome1, Oben, Odiferous, Openstrings, PhnxFyreG, Pnm, Public Menace, PublicassistantJr, Pugglewuggle, Quinxorin, Raysonho, Reaper Eternal, RedWordSmith, RenamedUser01302013, Rjwilmsi, Rob.bosch, Rocketrod1960, Ros Power, Rwxrwxrwx, Schumi555, Shell Kinney, Shorty114, Sjefskenneth, Stephan Leeds, Stryn, Tedickey, TenPoundHammer, TeslaRoadster, The Last Melon, TheParanoidOne, Titodutta, Tyler, Ukexpat, UncleDouggie, Urocyon, Vegaswikian, Wknight94, Woohookitty, Zabolzadeh, 206 anonymous edits

Image Sources, Licenses and Contributors

Image:Filemanager31.gif *Source*: http://en.wikipedia.org/w/index.php?title=File:Filemanager31.gif *License*: unknown *Contributors*: Bhowden, Codename Lisa, John of Reading, 1 anonymous edits

File:Fileman-OFM.png *Source*: http://en.wikipedia.org/w/index.php?title=File:Fileman-OFM.png *License*: unknown *Contributors*: User:Linuxgal

Image:Midnightcommander.png *Source*: http://en.wikipedia.org/w/index.php?title=File:Midnightcommander.png *License*: unknown *Contributors*: Createaccount

Image:GNUstep-liveCD.png *Source*: http://en.wikipedia.org/w/index.php?title=File:GNUstep-liveCD.png *License*: unknown *Contributors*: User:Gürkan Sengün

Image:Nautilus-spatial.png *Source*: http://en.wikipedia.org/w/index.php?title=File:Nautilus-spatial.png *License*: unknown *Contributors*: Norman Rogers

Image:FSV-OSX-screenshot.png *Source*: http://en.wikipedia.org/w/index.php?title=File:FSV-OSX-screenshot.png *License*: unknown *Contributors*: Software by Daniel Richard G.

File:Windows1.0.png *Source*: http://en.wikipedia.org/w/index.php?title=File:Windows1.0.png *License*: unknown *Contributors*: Aiyizo, Codename Lisa, Diego Moya, FleetCommand, Frogger3140, Gan Luo, Ghettoblaster, Happy Dude, Kubek15, Neurolysis, Nicolas Love, Remember the dot, Vanished user aoiowaiuyr894isdik43, Warren, 12 anonymous edits

File:Windows 3.0 workspace.png *Source*: http://en.wikipedia.org/w/index.php?title=File:Windows_3.0_workspace.png *License*: unknown *Contributors*: Chugun, Codename Lisa, Courcelles, Dancraggs, FSII, FleetCommand, Nicolas Love, Shlomital, Tyomitch, Warren, Yamla, 6 anonymous edits

File:Windows 95 Desktop screenshot.png *Source*: http://en.wikipedia.org/w/index.php?title=File:Windows_95_Desktop_screenshot.png *License*: unknown *Contributors*: AlistairMcMillan, Caypartis, Closedmouth, Codename Lisa, Damian Yerrick, Dan6hell66, Diego Moya, FleetCommand, Ghettoblaster, Koman90, Lordalpha1, McLoaf, Nicolas Love, Ruud Koot, Shlomital, Stefan2, ViperSnake151, Warren, 8 anonymous edits

Image:Microsoft Windows logo (Pre-XP).svg *Source*: http://en.wikipedia.org/w/index.php?title=File:Microsoft_Windows_logo_(Pre-XP).svg *License*: unknown *Contributors*: Alan, Codename Lisa, CommonsDelinker, Dharmadhyaksha, FleetCommand, Rezonansowy, Stas1995, Sémhur, Technical 13

Image:Windows logo - 2002.svg *Source*: http://en.wikipedia.org/w/index.php?title=File:Windows_logo_-_2002.svg *License*: unknown *Contributors*: User:FleetCommand

Image:Windows logo - 2012.svg *Source*: http://en.wikipedia.org/w/index.php?title=File:Windows_logo_-_2012.svg *License*: unknown *Contributors*: User:Fry1989

File:WindowsCE7.png *Source*: http://en.wikipedia.org/w/index.php?title=File:WindowsCE7.png *License*: unknown *Contributors*: Koman90, 1 anonymous edits

File:Windows Updated Family Tree.png *Source*: http://en.wikipedia.org/w/index.php?title=File:Windows_Updated_Family_Tree.png *License*: unknown *Contributors*: User:NOKIA 3120 classic

File:Magnify-clip.png *Source*: http://en.wikipedia.org/w/index.php?title=File:Magnify-clip.png *License*: unknown *Contributors*: User:Erasoft24

Image:Msdosad.jpg *Source*: http://en.wikipedia.org/w/index.php?title=File:Msdosad.jpg *License*: unknown *Contributors*: Bkell, Melesse, SchmuckyTheCat, Skier Dude, Swtpc6800, Tyomitch, 1 anonymous edits

File:US Navy 110129-N-7676W-152 Culinary Specialist 3rd Class John Smith uses the existing DOS-based food service management system aboard the aircraft .jpg *Source*: http://en.wikipedia.org/w/index.php?title=File:US_Navy_110129-N-7676W-152_Culinary_Specialist_3rd_Class_John_Smith_uses_the_existing_DOS-based_food_service_management_system_aboard_the_aircraft_.jpg *License*: unknown *Contributors*: Sandstein

File:Windows95BOXSHOT.png *Source*: http://en.wikipedia.org/w/index.php?title=File:Windows95BOXSHOT.png *License*: unknown *Contributors*: BWCNY, Closedmouth, Kahlil88, Koman90, Radon210, Stefan2

File:Windows 95 architecture.png *Source*: http://en.wikipedia.org/w/index.php?title=File:Windows_95_architecture.png *License*: unknown *Contributors*: User:Ruud Koot

File:Windows 95 at first run.jpg *Source*: http://en.wikipedia.org/w/index.php?title=File:Windows_95_at_first_run.jpg *License*: unknown *Contributors*: Codename Lisa, Diannaa, Hohho56oy, N. Harmonik, Ruud Koot, 1 anonymous edits

Image:Windows NT 4.0 SERVER CoverBOX.PNG *Source*: http://en.wikipedia.org/w/index.php?title=File:Windows_NT_4.0_SERVER_CoverBOX.PNG *License*: unknown *Contributors*: BWCNY, Przemub

Image:Windows NT WorkstationOS.PNG *Source*: http://en.wikipedia.org/w/index.php?title=File:Windows_NT_WorkstationOS.PNG *License*: unknown *Contributors*: BWCNY, Przemub

Image:Windows 2000 Explorer.png *Source*: http://en.wikipedia.org/w/index.php?title=File:Windows_2000_Explorer.png *License*: unknown *Contributors*: 6. клас, Bhowden, Codename Lisa, Kahlil88, Kozuch, RyanGerbil10, Tyomitch, Warren

Image:Windows Explorer XP.png *Source*: http://en.wikipedia.org/w/index.php?title=File:Windows_Explorer_XP.png *License*: unknown *Contributors*: AussieLegend, Diannaa, G&CP, Themodernizer

Image:Windows Explorer search puppy.png *Source*: http://en.wikipedia.org/w/index.php?title=File:Windows_Explorer_search_puppy.png *License*: unknown *Contributors*: Lifemaestro, Warren

Image:Thumbnailviewxp.png *Source*: http://en.wikipedia.org/w/index.php?title=File:Thumbnailviewxp.png *License*: unknown *Contributors*: Astroview120mm, AussieLegend, Kathleen.wright5, N. Harmonik, Warren

Image:Windows Explorer Vista.png *Source*: http://en.wikipedia.org/w/index.php?title=File:Windows_Explorer_Vista.png *License*: unknown *Contributors*: Drilnoth, FleetCommand, Sdrtirs, Skier Dude, SuperMarioMan, Themodernizer, 2 anonymous edits

Image:Details.PNG *Source*: http://en.wikipedia.org/w/index.php?title=File:Details.PNG *License*: unknown *Contributors*: Akhristov, Kozuch, Mardus, Timbernerslee, Warren

Image:Windows Explorer Windows 7.png *Source*: http://en.wikipedia.org/w/index.php?title=File:Windows_Explorer_Windows_7.png *License*: unknown *Contributors*: Calmer Waters, Codename Lisa, De728631, Fetchcomms, Himanis Das, Horiaeugen, Koman90, Lordalpha1, Melesse, RenamedUser01302013, Skier Dude, SuperMarioMan, Washington Irving Esquire, 1 anonymous edits

File:Windows Explorer Ribbon.png *Source*: http://en.wikipedia.org/w/index.php?title=File:Windows_Explorer_Ribbon.png *License*: unknown *Contributors*: Microsoft

MIX
Papier aus verantwortungsvollen Quellen
Paper from responsible sources
FSC® C105338

Printed by Books on Demand GmbH, Norderstedt / Germany